AIRBOARD TEC

CW00501012

BRISTOL FIGHTER F.2B.

(190 H.P. ROLLS-ROYCE.)

RIGGING NOTES.

BRISTOL FIGHTER F.2B.

⎡ (I). 190-H.P. ROLLS-ROYCE.* ⎤
⎢(II). 200-H.P. HISPANO SUIZA.†⎥

MANUFACTURERS' ORDER OF ERECTION.

1. Fuselage assembled and trued up.

2. Tanks fitted with Fuselage in Flying Position.

3. Lower Centre Section Plane and Undercarriage fitted.

4. Engine mounted with Machine in Flying Position.

5. Tail Skid fitted.

6. Tail Plane Actuating Gear fitted.

7. Fuselage covered and doped.

8 Upper Centre Section Plane fitted with Machine in Flying Position.

9. Main Planes attached with Machine in Flying Position.

10. Empennage fitted.

11. Controls connected up and adjusted.

---

* For Bristol Fighter F. 2A (190-H.P. Rolls-Royce) see **Appendix II.**
† For Bristol Fighter F. 2B (200-H.P. Hispano Suiza) see **Appendix I.**

BRISTOL FIGHTER F.2B.

(190-H.P. ROLLS-ROYCE.)

TRUING UP THE FUSELAGE (See Fig. 1).—The Fuselage is **symmetrical** throughout in plan view and from the top of Side Strut 4 to the horizontal Sternpost Tube in side elevation.

Insert a steel tube, to fit without play, into the horizontal Sternpost Tube and to project about one foot each side.

Mark the middle points of Side Struts 4, 5, 6, 7, 8, 9 and 10, on both sides.

Measure the *vertical* distance of the middle points of Side Struts 4 from the Bottom Longerons and mark points on Side Struts 1, 2 and 3, at this vertical distance from the Bottom Longerons.

Lightly clamp a straightedge transversely across Side Struts 1, the marked points on these Struts to be on the *upper edge.*

Stretch two lines, one each side from the *axis* of the horizontal Sternpost Tube, to the *upper edge* of the straightedge.

Mark the middle points of all Cross Struts, top and bottom.

Fit Jigs (See Fig. 2), to ensure that the correct distance is maintained between Sternpost Tube and Tail Plane Kingposts, over which Upper and Lower Fins fit.

Stretch another line, from the *axis* of the Kingpost for the Upper Fin to the *middle point* of the Front Top Cross Strut. This line will be *inside* the Fuselage.

By means of the Fuselage Tie Rods, true up, until:—

(*a*) All marked points on Side Struts are in the same horizontal plane as the two *outside* stretched lines.

Check by spirit level on the horizontal portions of the Top Longerons, longitudinally and transversely, and at each Side Strut by a try square, the base of the try square to be along the Strut, and the marked point to be on the *upper edge* of the Blade.

Bristol Fighter F.2B (190 H.P. Rolls-Royce).

The *outer* stretched line should touch the *upper edge* of the blade at each point checked.

(*b*) All marked points on Cross Struts and the *internal* stretched line be in the same vertical plane.

Check by dropping a plumb line from or to any marked point on a Cross Strut.

This plumb line should touch the *internal* stretched line.

TRUING UP UNDERCARRIAGE (See Fig. 3).—The Lower Centre Section Plane must be fitted *before* the Undercarriage can be bolted in position.

To true up the Undercarriage, adjust the Front Cross Bracing Wires until corresponding diagonals are equal.

PLACING MACHINE IN FLYING POSITION (See Fig. 4).—Before fitting and truing up the Upper Centre Section Plane and attaching the Main Planes it is necessary to get the Machine in Flying Position.

To do this, block the Machine up under the Undercarriage Struts and support the Tail on a trestle placed near the Sternpost Tube.

It is convenient to have a special trestle made to support the Machine under the Skidpost. (See Fig. 5).

Place blocks, sufficiently high to clear all obstacles, on the Gunner's Platform, ensuring that the blocks are accurately of the same height.

Place a straightedge across blocks and level longitudinally by raising or lowering Tail.

Level transversely by packing blocks under the Undercarriage Struts.

FITTING THE UPPER CENTRE SECTION.—The Upper Centre Section Struts are first bolted to the Upper Centre Section Plane, which is then lifted into position and the *lower ends* of the Struts bolted to the Centre Section Strut Fittings on the Top Longerons.

Cross Bracing Wires are then loosely connected.

The Main Plane Landing Wires are attached to the Lugs on the Upper Centre Section Plane.

TRUING UP CENTRE SECTION (See Fig. 4).—Before truing up Centre Section see that the Machine is in Flying Position.

The Lower Centre Section Plane cannot be adjusted in a *fore* and *aft* direction and is adjusted transversely by making the Front Cross Bracing Wires equal.

The Upper Centre Section Plane is adjusted longitudinally by the Side Cross Bracing Wires, and transversely by the Front Cross Bracing Wires.

True up Upper Centre Section Plane to be symmetrical transversely and to have a *Stagger* of 16.9″. Note.—In Machines Nos. A. 7177 and B. 1101 onwards the *Stagger* is 18.1″. (9th July, 17).

Bristol Fighter F.2B (190 H.P. Rolls-Royce).

Check by dropping plumb lines from the Leading Edge of Upper Centre Section Plane.

The *fore* and *aft* horizontal distance between the Leading Edge of Lower Centre Section Plane and the plumb lines should be 16.9″. [For Machines Nos. A. 7177 and B. 1101 onwards 18.1″. (9th July, '17.)]

TRUING UP MAIN PLANES (See Fig. 6).—The Main Planes are assembled with their Leading Edges on the ground (See Fig. 7).

All Interplane Struts are fitted and Flying and Incidence Wires loosely connected.

The Planes are then lifted into position and bolted to the Centre Section Planes, and the Landing Wires loosely connected.

The Drag Wires are also loosely connected.

To true up Main Planes drop plumb lines from the Leading Edges of the Upper Main Planes at six points, three on each side.

True up until :—

　*(a)　All plumb lines are in line when viewed from the side.

　(b)　The *Dihedral* is $3\frac{1}{2}°$ on Front and Rear Spars for both Upper and Lower Main Planes.

　Check by Dihedral Board, or straightedge along the Spars, and Abney level.

　(c)　The *Stagger* is 16.9″ throughout. NOTE.—In Machines Nos. A. 7177 & B. 1101 onwards the *Stagger* is 18.1″. (9th July, 17).

　(d)　The *Gap* is $64\frac{1}{2}″$ throughout.

　Check as in Centre Section.

　(e)　The *Incidence* is 1° 42′ (roundly one and three-quarter degrees) throughout for both Upper and Lower Main Planes.

　Check by straightedge and Abney level, taking care to place the straightedge from Leading Edge to Trailing Edge at *Ribs*.

　Check by holding a lath in a vertical position to touch the Leading Edge of the Upper Main Plane and by a levelled straightedge from Leading Edge of Lower Main Plane to vertical lath.

　The *Gap* can then be measured directly.

　Check for Main Planes being square with Fuselage by taking measurements between Bottom Sockets of Front Outer Struts and Rudderpost.

　These measurements should be the same on both sides.

FIXING THE EMPENNAGE (See Fig. 8).—Fit Upper and Lower Fins in position and secure the Lower Fin to Fixing Bolt on the Skidpost, and the Upper Fin to the Bolts on the Top Cross Strut of the Fuselage, above the Skidpost.

*On account of Dihedral the inner plumb lines will be slightly in front of outer plumb lines, but for practical purposes all may be regarded as being in line.

Bristol Fighter F.2B (190 H.P. Rolls-Royce).

Fit the Tail Plane in Position and loosely connect the Bracing **Wires** **from** Fins to Tail Plane.

True up until the Tail Plane Spars are horizontal and **Rudder** Hinges on Fins are vertically over one another.

Check the former by spirit level, and the latter by plumb line.

Check for Tail Plane being square with Fuselage by taking **measure-** ments from lateral extremities of Rear Spars to Gun Pivot on **Gunner's** **Platform.** These measurements should be the same on both sides.

Hinge the Rudder to the Fins by placing in Position, and inserting the Cotter Pins, not forgetting to secure the latter by *Split Pins.*

Hinge the Elevators to the Tail Planes in a similar manner.

CONTROLS.—With Machine in Flying Position, the Top Elevator Control Cables (*i.e.,* those attached to top end of Lever on Elevator) should be so adjusted that the *Pilot's Control Stick is vertical* when **the** Elevators are horizontal.

The Bottom Elevator Control Cables should then be adjusted **to** be fairly slack.

With the Pilot's Control Stick *vertical,* the Ailerons should *droop* $\frac{3}{4}$ in. and the Aileron Control Wires should be *fairly slack.*

With Rudder Bar symmetrical in Fuselage, the Rudder and Tail Skid should be square with Fuselage and point directly *fore* and *aft.*

TAIL PLANE ACTUATING GEAR (See Figs. 9, 10 and 11).—The Tail Plane is adjusted by a large Hand Lever, working on a quadrant, **on** the right-hand side of the Pilot's Seat. This Lever raises or lowers the Tail Plane Front Spar. When the handle of the Lever is *raised,* the Leading Edge of the Tail Plane is *depressed,* and *vice versa.*

The total movement of the Leading Edge of the Tail Plane should be about 2 in.

When the Leading Edge of the Tail Plane is *midway* between its highest and lowest positions it should be about $\frac{1}{2}$ in. to $\frac{3}{4}$ in. *below* the centre line of the Side of the Fuselage. In other words, when the Tail Plane (and also therefore the Hand Lever) are in their *mean positions* the Tail Plane should have a *negative* Incidence of about $\frac{1}{2}$ in. to $\frac{3}{4}$ in.

The Actuating Cables between the Hand Lever and the Bell Cranks should be kept fairly taut.

Care should be taken that the Turnbuckles on the Bell Crank Ends are so adjusted that the right and left hand Bell Cranks are *parallel* **to** one another, top to top, and bottom to bottom, otherwise there will be a tendency to cant the Tail Plane Front Spar.

BRISTOL FIGHTER F2B.
(190 H.P. ROLLS-ROYCE.)

LIST OF PRINCIPAL DIMENSIONS.

Span of Upper Main Planes Span of Lower ,, ,,	$39'\ 3''$
Chord of Upper Main Planes Chord of Lower ,, ,,	$5'\ 6''$
Incidence of Upper Main Planes Incidence of Lower ,, ,,	$1\frac{3}{4}°$
Dihedral of Upper Main Planes Dihedral of Lower ,, ,,	$3\frac{1}{2}°$
Stagger	$16\cdot9''$*
Gap	$6+\frac{1}{2}''$
Overall Length	$26'\ 2''$
Height	$8'\ 9''$
Incidence of Tail Plane with Hand Lever in Mean Position	$-\frac{1}{2}''$ to $-\frac{3}{4}''$
Droop of Ailerons, with Pilot's Control Stick Central	$\frac{3}{4}''$
Elevators (with Pilot's Control Stick Central)	Horizontal

WEIGHTS IN LBS.

Weight light, including Guns and Mountings complete	1700
Military Load (Ammunition)	150
Human Load	360
Fuel, Oil and Water (Tanks full)	440
TOTAL	2650

*In Machines Nos. A. 7177 & B. 1101 onwards the Stagger is 18.1". (9th July, 17).

POINTS TO OBSERVE WHEN OVER-HAULING MACHINE.

See that the Leading Edges of the Main Planes are symmetrical about centre line of machine.

Examine the Bracing Wires for length and taut-ness in the Centre Section, and see that the Split Pins are in position.

Check the Dihedral.

Check the Stagger.

Check the Incidence.

See that the Interplane Struts are straight.

Examine all Main Plane Bracing Wires for length and tautness, and see that all Split Pins are in position, and that all Turnbuckles on Cables are locked.

Examine all Controls, Control Pulleys and Cables, and see that they work freely, and that all Turn-buckles on Cables are locked.

Examine Tail Plane and see that it is set correctly and is square with Machine, and that all Tail Plane Bracing Wires are correct both as to tautness and length, and that all Split Pins are in position.

Examine Tail Plane Actuating Gear and see that it works freely.

Examine Rudder and Fins and see that they are set straight and square with Machine.

Measure the Droop of the Ailerons and Elevators.

Examine Undercarriage and Skid.

Examine Tank Mountings and Connections.

Examine Engine Mounting, Engine Controls, and Engine Accessories.

APPENDIX I.

BRISTOL FIGHTER F.2B.

(200 H.P. HISPANO-SUIZA.)

In the case of the 200 H.P. Hispano-Suiza Machine—

(a) The *Dihedral* is $3\frac{1}{2}°$ for the Front Spars only on account of "*Wash Out*" on both sides. Check by Abney level and Straightedge along the Front Spars only.

(b) The *Stagger* is 19.7″ throughout. Check by measuring the horizontal fore and aft distance between the Leading Edge of Lower Main Planes and plumb lines dropped from the Leading Edge of Upper Main Planes. These measurements should be 19.7″.

(c) The Incidence is—

      (i) 1°42′ at Centre Section.

      (ii) 1°24′ at Inner Interplane Struts on *both* sides.

      (iii) 1° 0′ at Outer Interplane Struts on *both* sides.

Check by Abney level and straightedge, placing the latter from Leading Edge to Trailing Edge at Ribs.

Proceed otherwise as in the 190 H.P. Rolls-Royce Machine.

APPENDIX II.

BRISTOL FIGHTER F. 2A.
(190-H.P. ROLLS-ROYCE.)

TRUING UP THE FUSELAGE.—In this Machine the Fuselage is symmetrical in both plan and side elevation.   Therefore in truing it up the middle points of all Cross Struts, top and bottom, should be in the same vertical plane, whilst the middle points on all Side Struts must be in the same horizontal plane.

In this Machine there is no Lower Centre Section Plane, and the Lower Main Planes are attached to a Wing Anchorage Frame which allows of lateral adjustment only.

Otherwise proceed exactly as in F. 2B.

BRISTOL FIGHTER F2.A.

(190-H.P. ROLLS-ROYCE).

LIST OF PRINCIPAL DIMENSIONS.

Span of Upper Main Planes Span of Lower ,, ,,	39ft. 3in.
Chord of Upper Main Planes Chord of Lower ,, ,,	5ft. 6in.
Incidence of Upper Main Planes Incidence of Lower ,, ,,	$1\frac{3}{4}$°.
Dihedral of Upper Main Planes Dihedral of Lower ,, ,,	$3\frac{1}{2}$°.
Stagger	17.1in.
Gap	64.5in.
Overall Length	26ft. 3in.
Height from ground to highest point of Machine	11ft. 2in.
Tail Span	12ft. 0in.
Incidence of Tail Plane with Hand Lever in Mean Position	$\frac{1}{2}$in.
Droop of Ailerons (with Pilot's Control Stick central)	$\frac{3}{4}$in.
Elevators (with Pilot's Control Stick central)	Horizontal.

Weights, in lbs.—

Power Unit Dry	874
Aeroplane	853
Military Load	180
Human Load	360
Fuel, Oil and Water	400
TOTAL	2,667

A.B.T.D., T5. 1/18. BRISTOL FIGHTER F2B (190 H.P. Rolls-Royce).

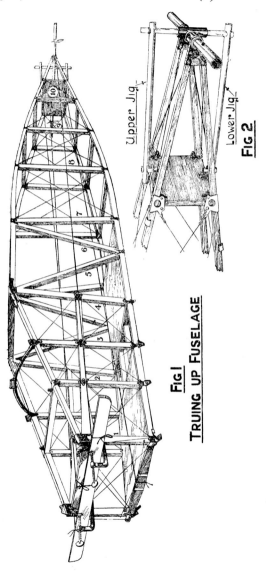

FIG 2

Upper Jig

Lower Jig

FIG I
TRUING UP FUSELAGE

A.B.T.D. T5. 1/18. BRISTOL FIGHTER F2B (190 H.P. Rolls-Royce).
FIG. 3.

A.B.T.D. T5. 1/18. BRISTOL FIGHTER F2B (190 H.P. Rolls-Royce).
FIG. 4.

A.B.T.D., T5. 1/18.   BRISTOL FIGHTER F.2B (190 H.P. Rolls-Royce).

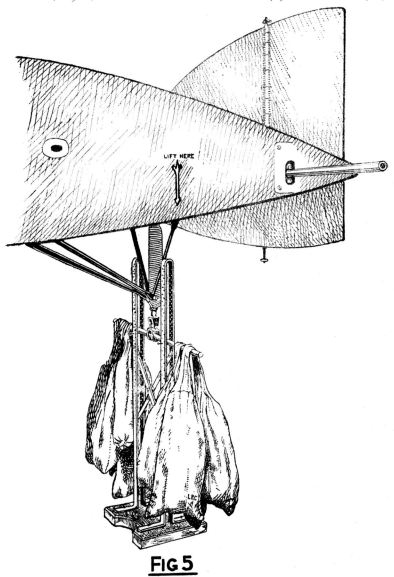

LIFT HERE

FIG 5

A.B.T.D., T.5. 1/18.　BRISTOL FIGHTER F2B (190 H.P. Rolls-Royce).

FIG. 6.

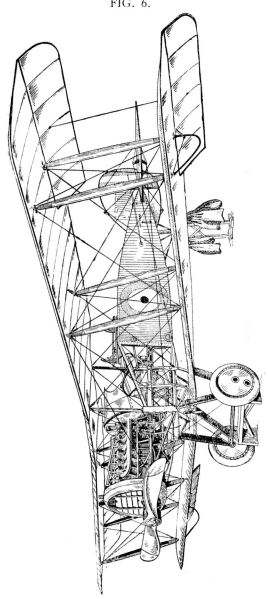

A.B.T.D. T5. 1/18. BRISTOL FIGHTER F2B (190 H.P. Rolls-Royce).
FIG. 7.

A.B.T.D. T5. 1/18. BRISTOL FIGHTER F2B (190 H.P. Rolls-Royce).
FIG 8.

A.B.T.D., T.5. 1/18.   BRISTOL FIGHTER F2B (190 H.P. Rolls-Royce)
FIG. 9.

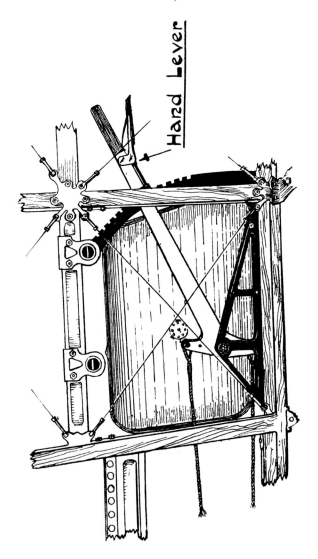

A.B.T.D., T5., 1/18.   BRISTOL FIGHTER F2B (190 H.P. Rolls-Royce)
FIG. 10.

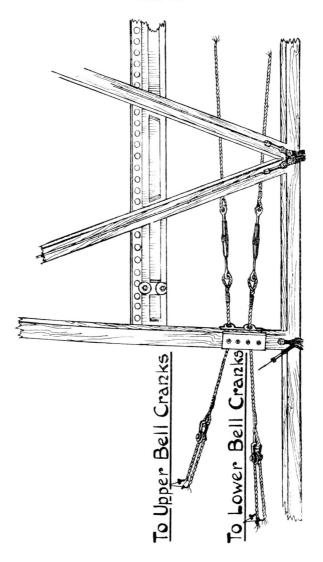

A.B.T.D., T5. 1/18.   BRISTOL FIGHTER F2B (190 H.P. Rolls-Royce).
FIG. 11.

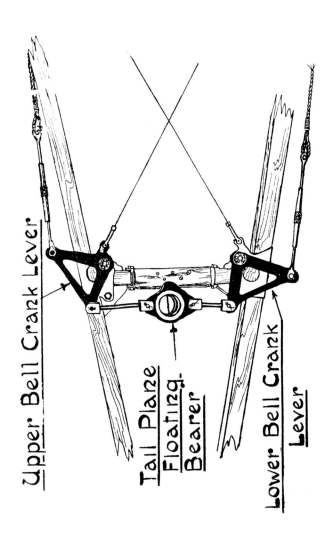

BRISTOL FIGHTER F.2B.

(i.) 190 H.P. ROLLS-ROYCE.   (ii.) 200 H.P. HISPANO-SUIZA.

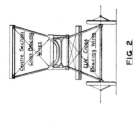

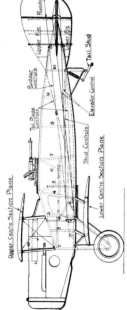

FIG. I.
SIDE ELEVATION
(LESS MAIN PLANES)

Flying Position.

The Machine is in Flying Position when the Gunner's Platform is level longitudinally and transversely.

Truing up the Fuselage.

Mark mid panels of Side Struts 4,5,6,7,8,9,10 and mark points on Side Struts 1,2, & 3 17.7" vertically above the bottom face of the Bottom Longerons. Stretch two lines, one on each side from marked points on Side Struts 1 to axis of horizontal Sternpost Tube. Mark middle points of all Cross Struts for a bottom. Stretch a centre line from mid point of Front Top Cross Strut to the axis of the King Post for the Upper Fin. Working from front to rear make Internal Cross Bracing Wires equal of each section and check by trammel. Make Top Cross Bracing Wires equal in each bay and similarly make Bottom Cross Bracing Wires equal in each bay by checking in each stage by trammel. True up Side Bracing Wires on one side until all marked points on Side Struts on that side are in reference to the outer stretched lines. A plumb line dropped from the mid point of a Top Cross Strut should strike the mid point of the corresponding Bottom Cross Strut and just touch the internal stretched line.

FIG 2
FRONT ELEVATION
(LESS MAIN PLANES)

Truing up Undercarriage.

Adjust Front Cross Bracing Wires making corresponding diagonals equal and check by trammel.

Truing up the Centre Section.

The Lower Centre Section Plane cannot be adjusted in a fore and aft direction and is adjusted transversely by making the Front Cross Bracing Wires equal. True up Upper Centre Section Plane to be symmetrical transversely and to have a Stagger of 16.9" (in later Machines A7177 onwards and B1101 onwards the Stagger will be 18.1")

Check by measuring the horizontal fore and aft distance between the Leading Edge of Lower Centre Section Plane and plumb lines dropped from the Leading Edge of Upper Centre Section Plane.

These measurements must be 16.9" (in later Machines A7177 onwards and B1101 onwards 18.1).

T5. 1/18.

BRISTOL FIGHTER F2.B.

(i.) 190 H.P. ROLLS-ROYCE.   (ii.) 200 H.P. HISPANO-SUIZA (see sheet 4).

FIG 3.

FRONT ELEVATION

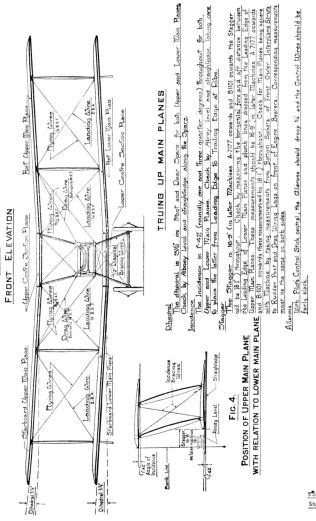

FIG. 4.

POSITION OF UPPER MAIN PLANE
WITH RELATION TO LOWER MAIN PLANE

TRUING UP MAIN PLANES

Dihedral
    The dihedral is 3½° on Front and Rear Spars for both Upper and Lower Main Planes. Check by Abney Level and straightedge along the Spars.

Incidence
    The incidence is 1°·42′ (roughly one and three-quarter degrees) throughout for both Upper and Lower Main Planes. Check by Abney Level and straightedge, taking care to place the latter from Leading Edge to Trailing Edge at Ribs.

Stagger
    The Stagger is 16·9″ (in later Machines A7177 onwards and B1101 onwards the Stagger will be 18·1″) throughout. Check by measuring the horizontal fore and aft distance between the Leading Edge of Upper Main Planes and plumb lines dropped from the Leading Edge of Upper Main Planes. These measurements should be 16·9″ in later Machines A7177 onwards and B1101 onwards these measurements will be 18·1″ throughout. Check for Main Planes being square with Machine by taking measurements from Bottom Sockets of Front Outer Interplane Struts to Rudder Post and Drag Wiring Lugs on front of Engine Bearers. Corresponding measurements must be the same on both sides.

Ailerons
    With Pilots Control Stick central, the Ailerons should droop ¾″ and the Control Wires should be fairly slack.

5 Sheet.
Sheet 2

BRISTOL FIGHTER F2B.

(i.) 190 H.P. ROLLS-ROYCE.   (ii) 200 H.P. HISPANO-SUIZA.

FIG 5.

GENERAL PLAN

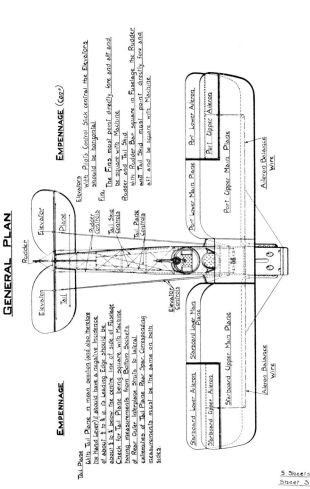

EMPENNAGE

Tail Plane

With Tail Plane in mean position (and also therefore the Hand Lever) it should have a negative Incidence of about $\frac{1}{4}$ to $\frac{1}{4}$ ie. its Leading Edge should be about $\frac{1}{4}$ to $\frac{1}{2}$ below the centre line of side of Fuselage. Check for Tail Plane being square with Machine by taking measurements from Bottom Sockets of Rear Outer Interplane Struts to lateral extremities of Tail Plane Rear Spar. Corresponding measurements must be the same on both sides.

EMPENNAGE (con.)

Elevators
with Pilot's Control Stick central the Elevators should be horizontal.

Fins The Fins must point directly fore and aft and be square with Machine.

Rudder and Tail Skid
with Rudder Bar square in Fuselage the Rudder and Tail Skid must point directly fore and aft and be square with Machine.

Starboard Lower Aileron
Starboard Upper Aileron
Starboard Lower Main Plane
Starboard Upper Main Plane
Aileron Balance wire

Port Lower Aileron
Port Upper Aileron
Port Lower Main Plane
Port Upper Main Plane
Aileron Balance wire

Rudder
Elevator
Plane
Tail
Elevation
Rudder Controls
Tail Skid Controls
Tail Plane Controls
Elevator Controls

T₅. 1/18.

BRISTOL FIGHTER F.2B.

(i.) 200 H.P. Hispano-Suiza.

Fig. 6

POSITION OF UPPER CENTRE SECTION PLANE WITH RELATION TO LOWER CENTRE SECTION PLANE.

Fig. 7

POSITION OF UPPER MAIN PLANE WITH RELATION TO LOWER MAIN PLANE.

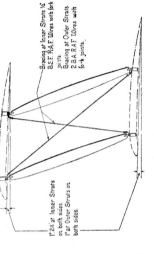

Bracing at Inner Struts ½" B.S.F. R.A.F. Wires with fork joints.
Bracing at Outer Struts 2 B.A. R.A.F. Wires with fork joints.

1"24' at Inner Struts on both sides. 1" at Outer Struts on both sides.

FOR 200 H.P. HISPANO SUIZA ONLY —— TRUING UP MAIN PLANES.

Dihedral. The Dihedral is 3½". Check by Abney Level and Straightedge on Front Spars only.

Stagger. The Stagger is 19.7 throughout. Check by measuring the horizontal fore & aft distance between Leading Edge of Lower Main Planes and plumb lines, dropped from Leading Edge of Upper Main Planes. These measurements must be 19.7 throughout.

Incidence. The Incidence is (i) 1° 42' at Centre Section (ii) 1° 24' at Inner Struts on both sides. (iii) 1° at Outer Interplane Struts on both sides. Check by Abney Level and Straightedge placing the latter from Leading Edge to Trailing Edge at Ribs.

Ailerons. With Pilots Control Stick central the Ailerons should droop ¾" and the Control Wires should be fairly slack. Check for Main Planes being square with machine by taking measurements from Bottom Sockets of Front Outer Interplane Struts to Rudder Post and Drag Wiring Lugs on Front Engine Bearers. Corresponding measurements must be the same on both sides.

BRISTOL FIGHTER F.2B

(i.) 190 H.P. ROLLS-ROYCE.   (ii.) 200 H.P. HISPANO-SUIZA

RAF. WIRES FOR ALL F 2 B. TYPES.

Part No	Description	Size	No Off	Drawing No	A. approx	B. approx	Remarks
					Lengths in inches		
1	Front Inner Flying	9/32 B.S.F.	4	A G S 349.	91.5	84	
2	Rear Inner Flying	9/32 B.S.F.	4	. .	91.5	84	
3	Front Outer Flying	1/4 B.S.F.	4	. 348.	111	103	
4	Rear Outer Flying	1/4 B.S.F.	4	. .	111	103	
5	Front Inner Landing	1/4 B.S.F.	2	. .	87	81	
6	Rear Inner Landing.	1/4 B.S.F	2	. .	87	81	
7	Front Outer Landing	2 B.A.	2	. 347	104	98	
8	Rear Outer Landing.	2 B.A.	2	. .	104	98	
9	Inner Bay Incidence (Long)	1/4 B.S.F.	2	. 322	76	70	
10	" " (Short)	1/4 B.S.F.	2	. .	60	53.5	
11	Outer Bay Incidence (Long)	2 B.A	2	. 321	76.5	71	
12	" " (Short)	2 B.A.	2	. .	60	54	
13	Top Centre Section Cross Bracing	4 B.S.F	2	F. 3382			Special
14	Bottom Centre Section Cross Bracing	4 B.S.F	2	AG S 322	30		
15	U/c Cross Bracing	9/32 B.S.F	2	. 349	50	44	
16	Aileron Connecting	4 B.A.	2	. 346	63	59	
17	Tail Plane Top Front	2 B.A.	2	. 321	54	50	
18	" Bottom Front	2 B.A.	2	. .	49	45	
19	" Top Rear	2 B.A.	2	. .	65.5	61.5	
20	" Bottom Rear	2 B.A.	2	. .	59.5	54.5	
21	Centre Section Side Bracing (Long)	1/4 B.S.F.	2	. 309			
22	" " (Short)	4 B.S.F.	2	. .			

BRISTOL FIGHTER F.2A. 190 H.P. ROLLS-ROYCE.

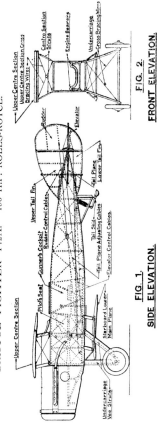

FIG. 1.
SIDE ELEVATION.
(LESS PORT MAIN PLANES)

FIG. 2.
FRONT ELEVATION.
(LESS MAIN PLANES)

Flying Position.

The Machine is in Flying Position when the Top Longerons in the Pilot's Cockpit are level longitudinally and transversely.

Truing up the Fuselage.

The Fuselage is symmetrical in Plan and Side Elevation. Mark the mid points of all Side Struts. Stretch two centre lines, one each side from mid points of Front Struts to axis of Horizontal Sternpost Tubes. Mark the mid points of all Cross Struts, top and bottom. Stretch another centre line from the mid point of the Front Top Cross Strut to the axis of the King post of the upper fin. Working from front to rear adjust internal Cross Bracing Wires making corresponding diagonals equal at each section and check by Tramrel all through. Adjust the Top Cross Bracing Wires until the mid points of all Top Cross Struts are in line with the top points of all Side Struts on that side are in line with outer stretched centre line. Adjust the side Bracing Wires on one side until the mid points of all Side Struts on that side are in line with outer stretched centre line. Proceed similarly for the other side. Adjust the Bottom Cross Bracing Wires until a plumb line dropped from the mid point of a Top Cross Strut strikes the mid point of its corresponding Bottom Cross Strut and just touches the internally stretched centre line.

Truing up the Under-carriage.

Adjust Front Cross Bracing Wires making corresponding diagonals equal and check by tramrel.

Truing up the Centre Section.

Adjust Front and Rear Cross Bracing Wires of Lower Wing Anchorage Frame until corresponding diagonals are equal and check by tramrel. Adjust the upper Centre Section Plane to be symmetrical about the vertical centre line of Machine. Adjust by Front Cross Bracing Wires making Upper Wires equal and Lower Wires equal. The Upper Centre Section Plane should be adjusted so that in Flying Position plumb lines dropped from its Leading Edge are 12.9 inches horizontally in front of the Front Edge of the front Top Cross Strut of Fuselage. Adjust by Side Bracing Wires.

BRISTOL FIGHTER F.2.A. 190 H.P. ROLLS-ROYCE.

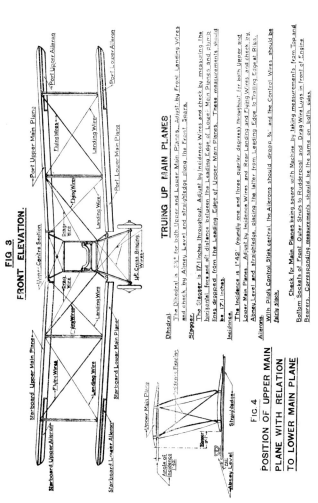

FIG 3
FRONT ELEVATION.

FIG 4
POSITION OF UPPER MAIN PLANE WITH RELATION TO LOWER MAIN PLANE

TRUING UP MAIN PLANES

Dihedral.

The Dihedral is 2¾° for both Upper and Lower Main Planes. Adjust by Front Landing Wires and check by Abney Level and straightedge along the Front Spars.

Stagger.

The Stagger is 17¼ inches throughout. Adjust by Incidence Wires and check by measuring the horizontal fore and aft distance between the Leading Edge of Lower Main Planes and plumb lines dropped from the Leading Edge of Upper Main Planes. These measurements should be 17¼ inches.

Incidence.

The Incidence is 1°42' (roughly one and three quarter degrees) throughout for both Upper and Lower Main Planes. Adjust by Incidence Wires and Rear Landing and Flying Wires and check by Abney Level and straightedge, placing the latter from Leading Edge to Trailing Edge at Ribs.

Ailerons.

With Pilot's Control Stick central, the Ailerons should droop ¾" and the Control Wires should be fairly slack.

Check for Main Planes being square with Machine by taking measurements from Top and Bottom Sockets of Front Outer Struts to Rudderpost and Drag Wire Lugs in front of Engine Bearers. Corresponding measurements should be the same on both sides.

BRISTOL FIGHTER F2.A. 190 H.P. ROLLS-ROYCE

FIG. 5.

GENERAL PLAN

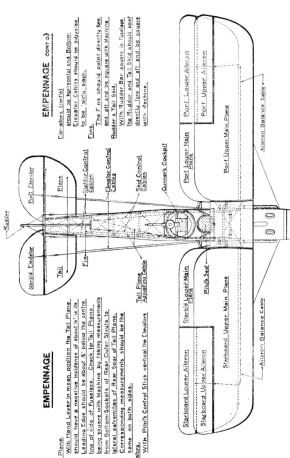

EMPENNAGE

Tail Plane

With Hand Lever in mean position the Tail Plane should have a negative incidence of about ¼ ie. its Leading Edge should be about ¾ below the centre line of side of Fuselage. Check for Tail Plane being square with Machine by taking measurements from Bottom Sockets of Rear Outer Struts to lateral extremities of Rear Spar of Tail Plane. Corresponding measurements should be the same on both sides.

Elevators.

With Pilot's Control Stick vertical the Elevators

EMPENNAGE (cont'd)

Elevators (cont'd)

should be horizontal and Bottom Elevator Cables should be adjusted to be fairly slack.

Fins.

The Fins should point directly fore and aft and be square with Machine.

Rudder & Tail Skid.

With Rudder Bar square in Fuselage the Rudder and Tail Skid should point directly fore and aft and be square with Machine.

4 Sheets,
Sheet 3.

T5. 1/18.

BRISTOL FIGHTER, F.2.A., 190 H.P. ROLLS-ROYCE.
RAF-WIRE LENGTHS.

Index letter (see diagram)	Description	Size		No off	Lengths A Ins.	B	Drawing No.	Remarks
A	Front & Rear Inner Flying Wires.	3/16	B.S.F.	8	91.5 Ins.	84 Ins.	A.C.S. 349	Wing External Bracing
B	" " " Landing Wire.	1/4	"	4	87 "	81 "	" 348	Ditto
C	" " Outer Flying Wires.	1/4	"	8	111 "	103 "	" 348	Ditto
D	" " " Landing Wire.	2	B.A.	4	104 "	98 "	" 347	Ditto
E	Inner Incidence Wire (long)	1/4	B.S.F.	2	76 "	70 "	" 348	Ditto
F	" " " (short)	1/4	"	2	60 "	53.5 "	" 348	Ditto
G	Outer " " (long)	2	B.A.	2	76.5 "	71 "	" 347	Ditto
H	" " " (short)	2	"	2	60 "	54 "	" 347	Ditto
I	Lower Centre Section Front & Rear CrossBracing Wire	1/4	B.S.F.	4	37.5 "	32 "	" 322	Bottom Wing Anchorage
J	Upper Centre Section Front Cross Bracing Wire (upper)	1/4	"	2	23 "	17.5 "	" 322	Top Centre Plane Attachment
K	" " " " (lower)	1/4	"	2	15.5 "	10 "	" 322	Ditto
L	Tail Plane. Top Front Bracing Wire.	2	B.A.	2	54 "	49 "	" 321	
M	" " Bottom " "	2	"	2	49 "	44 "	" 321	
N	" " Top Rear " "	2	"	2	66.5 "	60.5 "	" 321	
O	" " Bottom " "	2	"	2	59.5 "	54.5 "	" 321	
P	Undercarriage Cross Bracing Wire	3/16	B.S.F.	2	50 "	46 "	" 349	
Q	Aileron Control Wire.	2	B.A.	2	162 "	156 "	" 347	
R	" Connecting Wire	4	"	2	63 "	59 "	" 346	

NOTES.

De Havilland No. 5 (110 H.P. Le Rhone).

ATTACHING MAIN PLANES (See Fig. 4).

The Lower Main Planes are bolted in position and the Landing Wires loosely connected.

The Upper Main Planes are then lifted and bolted in position, the Interplane Struts fitted into their Sockets, and the Flying and Incidence Wires loosely connected.

The Gravity Tank is fitted on the Starboard Upper Main Plane. The centre line of Tank should be $9\frac{3}{8}''$ from the junction of the Planes.

TRUING UP MAIN PLANES (See Figs 4 and 5).

The Leading Edges of both Upper and Lower Main Planes should be symmetrical about centre line of Machine. Check by taking measurements from the Bottom Sockets of the Front Outer Struts to the Rudderpost.

These measurements should be the *same* on both sides.

Check for Leading Edges being parallel by placing straightedges across the Leading Edges of Upper and Lower Main Planes. Any *two* of these straightedges should be in line.

The *Dihedral* is $4\frac{1}{2}°$ for both Upper and Lower Main Planes. Check by straightedge along Front Spars and Abney level.

A line stretched over the Upper Main Planes from the *tops* of the Front Outer Struts should be $5\frac{3}{4}''$ vertically over the Upper Centre Section Plane. Check at lateral extremities of Upper Centre Section Plane.

The *Stagger* is the same as that of the Centre Section throughout.

The *Incidence* of the Starboard Main Planes should be 2° throughout. Check by straightedge and Abney level as in Centre Section.

On account of Propeller Torque the *Incidence* of the Port Main Planes is $2\frac{1}{4}°$ at the Outer Struts. Check as in Starboard Main Planes

FIXING THE EMPENNAGE (See Figs. 7 and 8).

The Tail Plane should be set in the third hole from the bottom. Any slight adjustment which it may require can be made after tests.

The Tail Plane must be level transversely, and square with the Fuselage. Check the former by spirit level along the Spars, and the latter by taking measurements from the Bottom Sockets of Front Outer Struts to the lateral extremities of the Rear Spar of the Tail Plane. These measurements should be the same on both sides.

The Fin should be set square with Fuselage and point directly *fore* and *aft*.

Hinge the Rudder to the Fin and Sternpost by placing in position and inserting the Split Pins.

Similarly hinge the Elevators to the Tail Plane.

CONTROLS.

With the Pilot's Control Stick central the *Droop* of the Ailerons is $\frac{3}{4}''$, and the Elevators are in direct continuation of the Tail Plane.

The Rudder and Tail Skid should point directly *fore* and *aft*, and be square with Machine when the Rudder Control Bar is central.

De Havilland No. 5 (110 H.P. Le Rhone).

(c) .djust the Cross Bracing Wires on *one* side until the middle points of all Side Struts in Rear Portion, and the lower marked point on the Front Strut are in line.
Check by try square at each middle point with reference to the *outer* stretched line.

(d) Place a straightedge transversely across the Top Longerons at the point where the Tail Skid is fitted. Place another straightedge transversely across Top Longerons at points about 2½" from Strut Sockets.
By means of the Side Cross Bracing Wires not yet tensioned true up until the upper edge of the second straightedge at each point is in line with the upper edge of the first straightedge.

(e) Tension the Bottom Cross Bracing Wires.
The Fuselage should now be completely true.
Check finally by try square method in relation to all three lines.

TRUING UP THE UNDERCARRIAGE (See Figs. 4 and 6).

Adjust the Front Cross Bracing Wires until corresponding diagonals are equal. Check by trammel.

PLACING MACHINE IN FLYING POSITION.

Before truing up the Centre Section and attaching the Main Planes it is necessary to get the Machine in Flying Position. To do this, block the Machine up under the Undercarriage Struts and support the Tail on a Trestle placed near the Tail Skid.

The Machine is in Flying Position when it is level transversely and when the angle of *Incidence* of the Lower Centre Section Planes is 2°.

Level transversely by straightedge and spirit level across the Top Longerons and make any adjustments by packing blocks under the Undercarriage Struts.

Make longitudinal adjustments by raising or lowering the Ta unti the *Incidence* of the Lower Centre Section Planes is 2°. Check by straightedge and Abney level, placing the straightedge from Leadin Edge to Trailing Edge at *Ribs*.

TRUING UP CENTRE SECTION (See Figs. 3 and 6).

By means of the Centre Section Anti-Drag Wires true until the Upper Centre Section Plane is symmetrical about the vertica centre line of the Machine. Check by trammel, making corresponding diagonals between Upper Centre Section Plane and Top Longerons of Fuselage equal.

By means of the Centre Section Side Bracing Wires and Anti-Drag Wires adjust the Upper Centre Section Plane longitudinally until a plumb line dropped from a point 2" in *rear* of centre line of Front Spar of Upper Centre Section Plane strikes a point 3½" in *rear* of the centre line of the Rear Spar of the Lower Centre Section Plane.

The *Incidence* of the Upper Centre Section Plane should be 2 throughout. Check by Abney level and placing the latter from the Leading Edge to Trailing Edge at Ribs.

DE HAVILLAND No. 5.

[110-H.P. LE RHONE]

TRUING UP FUSELAGE (See Figs. 1 & 2).

The Fuselage is in two portions (1) The Front Portion, (2) The Rear Portion. These are joined by *Butt Joints* and *Fish Plates*. The top joints are $3\frac{3}{4}''$ from the centres of bolts through Top Longerons of the Front Portion vertically over the last Side Struts. The bottom joints are immediately in rear of the last Side Struts of the Front Portion.

The Rear Portion is symmetrical both in plan view and side elevation; the Front Portion is symmetrical in plan view only. To true up, mark the middle points of all Struts of the Rear Portion.

Mark the middle points of the Front Struts of the Front Portion and mark a second set of points on these Struts 4" *below* the middle points.

Lightly clamp a straightedge transversely across the middle points of the Struts where the Tail Skid is fitted, the marked points to be on the *upper edge.*

Lightly clamp another straightedge transversely across the Front Struts, the *lower marked* points to be on the *upper edge* of the straightedge.

Stretch two lines, one on each side of the Fuselage, from the upper edge of *rear* straightedge to upper edge of *front* straightedge.

Stretch a centre line above the Fuselage from the *axis* of the Rudderpost to the middle point of the Front Top Cross Strut.

Work from the junction of Front and Rear Portions and

(a) Adjust the Internal Cross Bracing Wires until diagonals are equal at each section. Check by trammel.

(b) Adjust the Top Cross Bracing Wires until the middle points of all Top Cross Struts are in the *same* straight line. Check by try square at each middle point ; the base of the try square should be along the Strut and the middle point of the Strut should be on the edge of the blade ; the *top centre line* should just touch the *same edge* of the blade.

NOTE :—The "Figs." refer to the Illustrations and *not* to the Rigging Diagrams.

DE HAVILLAND No. 5.

(110 H.P. LE RHONE.)

MANUFACTURERS ORDER OF ERECTION.

1. Front Portion of Fuselage assembled and trued up.

2. Rear Portion of Fuselage assembled and trued up.

3. Front and Rear Portions of Fuselage joined and lined up.

4. Tanks fitted.

5. Undercarriage fitted with Fuselage in inverted position.

6. Fairing fitted.

7. Engine mounted.

8. Instrument Board fitted.

9. Upper Centre Section Plane fitted with Machine in Flying Position and Lower Centre Section Planes covered and doped.

10. Gravity Tank fitted to Upper Main Planes.

11. Main Planes attached and trued up.

12. Engine Cowl fitted.

13. Empennage fitted.

14. Controls loosely connected.

15. Fuselage covered and doped.

16. Controls adjusted.

AIRBOARD TECHNICAL NOTES

DE HAVILLAND No. 5.

(110 H.P. LE RHONE.)

RIGGING NOTES.

DE HAVILLAND No. 5.

(110-H.P. LE RHONE.)

LIST OF PRINCIPAL DIMENSIONS.

Span of Upper Main Planes Span of Lower ,, ,,	25′8″ (Approx.)
Chord of Upper Main Planes Chord of Lower ,, ,,	4′6″
Incidence of Upper Main Planes Incidence of Lower ,, ,,	2°
Overall Length	22′0″ (Approx.)
Height	9′0″ ,,
Gap	5′0″

Stagger. A plumb line dropped from a point 2″ in *rear* of centre line of Front Spar of Upper Centre Section Plane must strike a point $3\frac{1}{2}''$ in *rear* of centre line of Rear Spar of Lower Centre Section Plane.

Dihedral (for both Upper and Lower Main Planes) - - - - - $4\frac{1}{2}°$

Allowance for Propeller Torque. Incidence at Port Outer Interplane Struts - $2\frac{1}{4}°$

Incidence of Tail Plane.
Tail Plane set in the third hole from the bottom.

Droop of Ailerons - - - - - $\frac{3}{4}''$

Droop of Elevator - - - - - Nil.

POINTS TO OBSERVE WHEN OVER-HAULING MACHINE

See that the Leading Edges of the Main Planes are symmetrical about centre line of Machine.

Examine the Bracing Wires for length and tautness in the Centre Section, and see that the Split Pins are in position and that all Lock Nuts are tight.

Check the Dihedral.

Check the Stagger.

Check the Incidence.

See that the Interplane Struts are straight.

Examine all Main Plane Bracing Wires for length and tautness, and see that all Split Pins are in position, and that all Lock Nuts are tight.

Examine all Controls, Control Pulleys and Cables, and see that they work freely and that Turnbuckles on Cables are locked.

Examine Tail Plane and see that it is set correctly, and is square with Machine, and that all Tail Plane Bracing Wires are correct both as to tautness and length, and that all Split Pins are in position and that all Lock Nuts are tight.

Examine Rudder and Fin and see that they are set straight and square with Machine.

Measure the Droop of the Ailerons and Elevators.

Examine Undercarriage and Skid.

Examine Tank Mountings and Connections.

Examine Engine Mounting, Engine Controls, and Engine Accessories.

A.B.T.D.T.5, 1/18. DE HAVILLAND No. 5. (110 H.P. Le Rhone). Fig. 1.

8

A.B.T.D.T5., 1/18.   De Havilland No. 5. (110 H.P. Le Rhone).   Fig. 3.

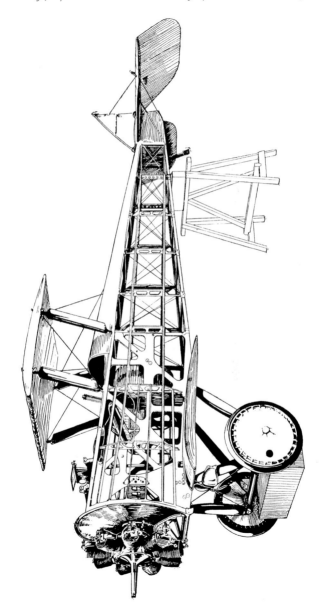

A.B.T.D.T5., 1/18.   De Havilland No. 5. (110 H.P. Le Rhone).   Fig. 4.

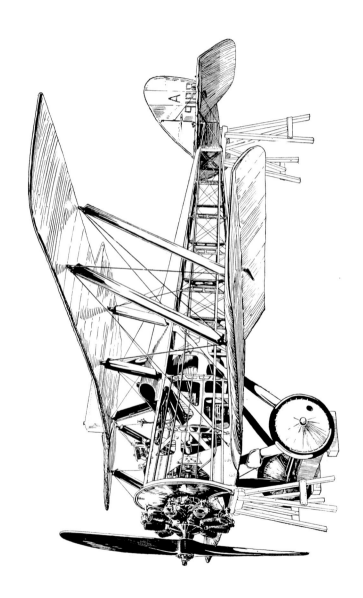

A.B.T.D.T5., 1/18. De Havilland, No. 5. (110 H.P. Le Rhone). Fig. 5.

A.B.T.D.T5., 1/18. DE HAVILLAND No. 5. (110 H.P. Le Rhone). Fig. 6.

A.B.T.D.T5., 1/18. De Havilland No. 5. (110 H.P. Le Rhone). Fig. 7.

A.B.T.D.T5., 1/18. DE HAVILLAND No. 5. (110 H.P. Le Rhone). Fig. 8.

AIRBOARD TECHNICAL NOTES

DE HAVILLAND No. 5.

(110 H.P. LE RHONE.)

RIGGING DIAGRAMS.

AEROPLANE DE HAVILAND Nº 5.
110 HP. LE RHONE.

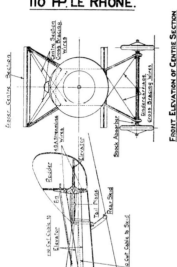

SIDE ELEVATION OF MACHINE.
Fig 1
TRUING UP

FRONT ELEVATION OF CENTRE SECTION
Fig 2
TRUING UP

Flying Position.

To get Machine in flying position, level transversely and adjust longitudinally until the incidence of the Lower Centre Planes is 2° Check by Straightedge and Abney level.

Truing up Fuselage.

Tension all internal Cross Bracing Wires until corresponding diagonals are equal throughout. By means of Top Cross Bracing wires true up until all mid points of Top Cross Struts are in one Straight line. By means of Side Cross Bracing wires (on one side) true up until mid points of Struts of Rear portion on that side are in line and a point 4 below mid point of Front Strut is in the same line. Place a Straightedge (ending board) across the Top Longerons just in front of second bay from the Tail. By means of the Side Cross Bracing Wires on the other side true up until the upper edge of another Straightedge, placed transversely across the Top Longerons at any point is in line with the Upper edge of the first Straightedge. Tension the Bottom Cross Bracing Wires.

Undercarriage.

True up undercarriage by adjusting Front Cross Bracing Wires until corresponding diagonals are equal. Check by trammel.

Centre Section.

True up by means of Centre Section Diagonal Bracing Wires. True up until Centre Section is symmetrical about Vertical Centre Line of machine and Check by trammel. By means of the Side Bracing Wires true up until the Stagger is as shown in Fig 4 and the incidence of Upper & Lower Centre Section Planes is 2° Check as shown in Fig 4.

AEROPLANE DE HAVILAND Nº 5.
110 H.P. LE RHONE.

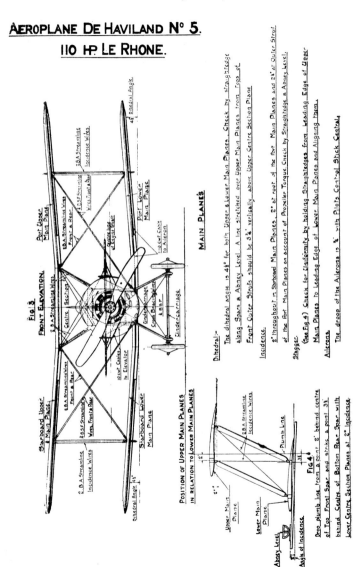

Fig 3
FRONT ELEVATION.

POSITION OF UPPER MAIN PLANES
IN RELATION TO LOWER MAIN PLANES
Fig 4.

Drop plumb line from a point 2" behind centre of Top Front Spar and strike a point 3½" behind Centre of Bottom Rear Spar with Lower Centre Section Planes at 2° incidence.

MAIN PLANES

Dihedral:—

The dihedral angle is 4½° for both Upper & Lower Main Planes. Check by straightedge along Spars & Abney Level. A line stretched over Upper Main Planes from Tops of Front Outer Struts should be 3¼" vertically above Upper Centre Section Plane.

Incidence.

2° throughout in Starboard Main Planes. 2° at root of the Port Main Planes and 2½° at Outer Strut of the Port Main Planes on account of Propeller Torque. Check by Straightedge & Abney Level.

Stagger.

(See Fig 4) Check for Uniformity by holding Straightedges from Leading Edge of Upper Main Planes to Leading Edge of Lower Main Planes and Aligning them.

Ailerons.

The droop of the Ailerons is ⅜" with Pilots Control Stick Central.

AEROPLANE DE HAVILAND Nº 5.

110 HP LE RHONE.

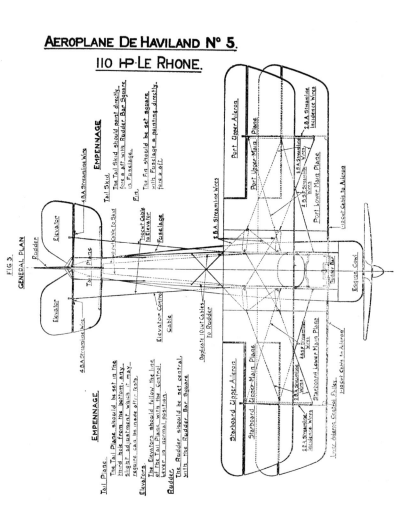

FIG 5
GENERAL PLAN

EMPENNAGE

Tail Skid.
The Tail Skid should point directly fore & aft with Rudder Bar Square in Fuselage.

Fin.
The Fin should be set square with Fuselage & pointing directly fore & aft.

EMPENNAGE

Tail Plane.
The Tail Plane should be set in the third hole from the bottom, any slight adjustment which it may require can be made after tests.

Elevators.
The Elevators should follow the line of the Tail Plane, with the Control Lever in normal position.

Rudder.
The Rudder should be set central with the Rudder Bar Square.

AEROPLANE DE HAVILLAND No. 5

(110 H.P. LE RHONE.)

RAF-WIRE LENGTHS.

No. per Machine	Position on Machine	Size of Screw	Overall Length A	Length between Screws B	Length of L.H. Screw	Length of R.H. Screw	Identification Mark
4	Flying Wires	$\frac{1}{4}$ B.S.F.	8′ $9\frac{5}{8}$″	8′ $5\frac{1}{4}$″	$1\frac{5}{16}$″	$3\frac{1}{16}$″	D.H. 5 A.
4	Landing Wires	2 B.A.	7′ $3\frac{7}{8}$″	6′ $11\frac{7}{8}$″	$1\frac{1}{8}$″	$2\frac{7}{8}$″	D.H. 5 B.
2	Incidence Wires (short)	2 B.A.	4′ 8″	4′ $4\frac{3}{4}$″	$1\frac{1}{8}$″	$2\frac{1}{8}$″	D.H. 5 C.
2	Incidence Wires (long)	2 B.A.	6′ $1\frac{3}{4}$″	5′ $10\frac{1}{2}$″	$1\frac{1}{8}$″	$2\frac{1}{8}$″	D.H. 5 D.
2	Centre Section Diagonal X Bracing	2 B.A.	4′ $5\frac{1}{8}$″	4′ $1\frac{7}{8}$″	$1\frac{1}{8}$″	$2\frac{1}{8}$″	D.H. 5 E.
2	Centre Section Side Bracing Wire	2 B.A.	2′ $10\frac{5}{8}$″	2′ $7\frac{3}{8}$″	$1\frac{1}{8}$″	$2\frac{1}{8}$″	D.H. 5 F.
2	Aileron Gap Wires	4 B.A.	5′ 3″	5′ 0″	1″	2″	D.H. 5 G.
2	Undercarriage X Bracing Wires	$\frac{1}{4}$ B.S.F.	3′ $4\frac{5}{8}$″	3′ 1″	$1\frac{5}{16}$″	$2\frac{5}{16}$″	D.H. 5 H.
2	Tail Plane Bracing Wires (top)	4 B.A.	2′ $3\frac{1}{2}$″	2′ $0\frac{1}{2}$″	1″	2″	D.H. 5 J.
2	Tail Plane Bracing Wires (bottom)	4 B.A.	2′ $1\frac{1}{4}$″	1′ $10\frac{1}{4}$″	1″	2″	D.H. 5 K.
2	Anti-Drag Wires	2 B.A.	5′ $3\frac{5}{8}$″	5′ $0\frac{3}{8}$″	$1\frac{1}{8}$″	$2\frac{1}{8}$″	D.H. 5 L.

4 Sheets Sheet 4.

NOTES.

AIRBOARD TECHNICAL NOTES

MAURICE FARMAN "SHORTHORN" BIPLANE [TYPE 1914], (80 H.P. RENAULT.)

RIGGING NOTES.

MAURICE FARMAN "SHORTHORN" BIPLANE (TYPE 1914).

(80 H.P. RENAULT.)

MANUFACTURERS' ORDER OF ERECTION.

1. Main Cell assembled and Centre Section trued up and Leading Edges of Main Planes adjusted to be straight.

2. Nacelle fitted.

3. Undercarriage fitted and trued up.

4. Tail Booms and Empennage fitted and trued up.

5. Main Cell completely trued up with machine in either "Rigging Position" or "Flying Position."

6. Empennage Trued up with Machine in Flying Position.

7. Controls connected up and adjusted.

8. Engine fitted, all Tank Connections made and Engine Controls connected.

MAURICE FARMAN " SHORTHORN "
BIPLANE [TYPE 1914]
(80 H.P. RENAULT.)

ERECTION AND PRELIMINARY TRUING UP MAIN CELL.

Interplane Struts are lettered as follows :—

Rear of Main Cell.

| T | S | R | P | N | M | L | K |
| A | B | C | D | E | F | G | H |

Front of Machine.

Erect Lower Main Plane, less Ailerons, on low trestles placed *under* Sockets for Struts A, T, C, R, F, M, H, K.

Bolt up the Upper Main Plane *complete* with Ailerons and Extensions and fit Struts A, T, C, R, F, M, H, K, into their *Top Sockets*.

By means of these Struts lift the Upper Main Plane bodily into position, fitting the *bottom ends* of the Struts into their Sockets on the Lower Main Plane.

Fit all remaining Struts and loosely connect all Main Cell Wires.

Connect four temporary Diagonal Wires, two in front and two in rear at Struts D, E, N, P.

TRUING UP THE CENTRE SECTION (See Fig. 1).

True up the Centre Section until all Centre Section Struts are vertical and the line joining Leading Edges of Upper and Lower Main Planes is at right angles to the *full chord* of the Centre Section, just *outside* Struts C, R, and F, M.

Check by means of a large square.

Stretch a line from the tops of the Bottom Struts Sockets at C. and F. This line must be 5 mm. below the *tops* of the Sockets at the *bottom* of Struts D. and E.

Next, by means of the Front Interplane Bracing Wires, adjust so that the Leading Edge of the Lower Main Planes is straight in plan view, and all Front Interplane Struts are straight and parallel.

Check by stretching a line from the *tops* of Bottom Sockets at Struts A. and H. This line should cut the *tops* of Bottom Sockets at Struts B, C, F and G, and be 5 mm. below the *tops* of Bottom Sockets at Struts D. and E.

Check for Struts being straight and parallel by aligning them from the side and viewing them from the front.

FITTING THE NACELLE.

To fit the Nacelle tie ropes *over* the Nacelle supports on Struts D, E, N, P, to the *bottom* of Struts C, F, M, R, respectively, thus slightly expanding the middle Bay of the Centre Section to allow the Nacelle to be placed in position without damaging the Struts D, E, N, P.

Lift Nacelle into position and bolt up to Struts D, E, N, P, and to Cross Tubes front and rear.

The distance from Leading Edges of Rear Centre Section Struts to centre line of Rear Engine Bearer Supports measured along the *top* of the Bottom Longerons should be 25 cms.

Maurice Farman " *Shorthorn* " *(30 H.P. Renault).*

PLACING MAIN CELL IN RIGGING POSITION.

Before fitting the Undercarriage it is necessary to get the Main Cell in Rigging Position. To do this place the Main Cell on high trestles placed under Struts A, T, and H, K.

Block up the supports until the Centre Section Struts are vertical, Check by plumb line.

TRUING UP THE UNDERCARRIAGE.

Adjust the Undercarriage Bracing Wires until the Rear Undercarriage Struts are vertical and in line with Interplane Struts R. and M. respectively, and all Undercarriage Struts on each side are in line with their Rear Strut, viewed from the front.

TAIL BOOMS AND EMPENNAGE.

The Tail Booms and Empennage are all assembled away from the Main Cell and the whole lifted into position, and the *front ends* of the Tail Booms bolted into the Sockets on the Rear Spars of the Centre Section Planes. The Tail is supported on high trestles placed under the Adjustable Steel Posts.

The Tail Booms should be straight and square with the Main Cell, and all Interboom Struts should be straight and parallel to Main Cell Struts.

Check the former by measuring the distance from any point on the rear of the Tail Booms to *bottom* Socket on Outermost Interplane Strut. This measurement should be the same on both sides; adjust for the latter by making diagonals equal in each bay of the Booms and check by *aligning* the Interboom Struts.

TRUING UP THE MAIN PLANES.

After fitting the Tail Booms and Empennage, place the Machine in Rigging Position, the supports being under the Undercarriage Main Skids and under the Adjustable Steel Posts. Adjust longitudinally by raising or lowering Tail, and transversely by packing blocks under the Undercarriage Main Skids.

Adjust the supports until the Centre Section Struts are vertical, and check for the latter by plumb lines.

In this position the Rear Spar of the Centre Section should be 40 mm. above the horizontal line, touching the Leading Edge of the Centre Section Plane. Check by a straightedge touching the Leading Edge of the Centre Section, bring to the horizontal and measure the height of the Rear Spar above this straightedge.

By means of the Incidence Wires and Rear Flying and Landing Wires adjust the Main Planes, working from the Centre Section outwards, so that this measurement is 36 mm. under Struts L. and G., and 32mm. under Struts H. and K., to allow for " *Wash In* " on the Port Side.

Under Struts B. and S. this measurement should be 44 mm., and under Struts A. and T., 48 mm., to allow for " *Wash Out* " on Starboard Side

Maurice Farman " Shorthorn " (80 H.P. Renault).

PLACING MACHINE IN FLYING POSITION.

The Machine is in Flying Position when the Engine Bearers and Top Longerons are horizontal and level transversely.

Block the Machine up as for " Rigging Position " and check for longitudinal and transverse level by placing a spirit level over the Engine Bearers.

It is more usual in workshop practice to get the Machine in Flying Position by setting the Main Cell so that the Main Planes have the correct *Incidence*, the procedure being as follows :—

Support the Machine as in Rigging Position, but set the Main Cell so that the Centre Section is level transversely, and so that it is at an *Incidence* of 65 mm. (the height of the Front Spar above the Rear Spar). Check by a levelled straightedge with one end under the centre of the Rear Spar and by direct measurement to under-surface of Front Spar.

NOTE.

The Main Planes can be trued up with the Centre Section in Flying Position, as follows :—

Adjust the Incidence and Rear Flying and Landing Wires until the *Incidence* at Struts L. and G. is 69 mm., and at Struts H. and K. 73 mm., to allow for " *Wash In* " on the Port Side ; and the Incidence at Struts B. and S. is 61 mm., and at Struts A. and T. 57 mm., to allow for " *Wash Out* " on Starboard Side.

TAIL PLANE.

The Tail Plane can be adjusted by means of four nuts, two each side, working on threads on each Adjustable Steel Post, which passes through a fitting on the Front Spar of Tail Plane.

To set the Tail Plane correctly the Machine must be in Flying Position.

By means of the Adjusting Nuts on the Threaded Adjustable Steel Posts adjust the Tail Plane to have an *Incidence* of 43 mm. Check by placing a straightedge with one end on the *under side* of the Rear Spar, bring the straightedge to the horizontal and adjust until the height of the Front Spar, above the straightedge, is 43 mm.

NOTE.

With 70 H.P. Renault this measurement should be 40 mm.

FINS.

The Fins should be square with Machine and point directly *fore* and *aft* with Machine in Flying Position.

CONTROLS.

With the Pilot's Control Stick central the *Droop* of the Ailerons is 25 mm., and the Elevator is in continuation of the Tail Plane.

The Rudders should be vertical and point directly *fore* and *aft* with Machine in Flying Position and Rudder Control Pedals symmetrical in Nacelle.

MAURICE FARMAN " SHORTHORN "
BIPLANE [TYPE 1914].
(80 H.P. RENAULT.)

LIST OF PRINCIPAL DIMENSIONS.

Span of Upper Main Planes   -   -   -   15m. 776mm.

Chord of ,,   ,,   ,,   -   -   -   2m. 013mm.

Angle of Incidence of Upper Main Planes -   -   4° 20'.

Span of Lower Main Planes   -   -   -   11m. 764mm.

Chord of ,,   ,,   ,,   -   -   -   2m. 013mm.

Angle of Incidence of Lower Main Planes -   -   4° 20'.

Gap -   -   -   -   -   -   -   -   1m. 900mm.

Stagger—Line joining Leading Edges of Upper and Lower
    Main Planes at right angles to full chord at Centre Section.

Dihedral -   -   -   -   -   -   -   - Nil.

" Wash in " at Port Outermost Struts-   -   -   8mm.

" Wash out " at Starboard Outermost Struts   -   8mm.

Overall Length -   -   -   -   -   9m. 300mm.

Height   -   -   -   -   -   -   3m. 150mm.

Span of Tail Plane -   -   -   -   5m. 490mm.

Chord of Tail Plane -   -   -   -   1m. 500mm.

Incidence of Tail Plane   -   -   -   -   43mm.

Diameter of Propeller   -   -   -   2m. 900mm.

Distance between Tail Booms   -   -   3m. 400mm.

MAURICE FARMAN "SHORTHORN"

BIPLANE [TYPE 1914].

(80 H.P. RENAULT).

POINTS TO OBSERVE WHEN OVERHAULING MACHINE.

See that the Leading Edges of the Main Planes are straight in plan view.

Examine the Bracing Wires in the Main Cell and Tail Booms for length and tautness, and see that they are not bent, and are correctly secured to the Wiring Lugs.

Check the Incidence.

See that all Interplane Struts are straight and are parallel to one another.

Examine all Controls, Control Pulleys and Cables and see that they work freely and that Turnbuckles on Cables are *locked*.

Examine Tail Plane and see that it is set correctly and is square with the Machine and that all Tail Plane Bracing Wires are correct, both as to length and tautness, and that they are properly secured to the Wiring Lugs.

Examine Rudders and Fins and see that they are set straight and square with Machine.

Measure the Droop of the Ailerons and Elevator, and see that the Hinges are secure.

Examine Undercarriage and Skid.

Examine Tank Mountings and Connections.

Examine Engine Mounting, Engine Controls, and Engine Accessories.

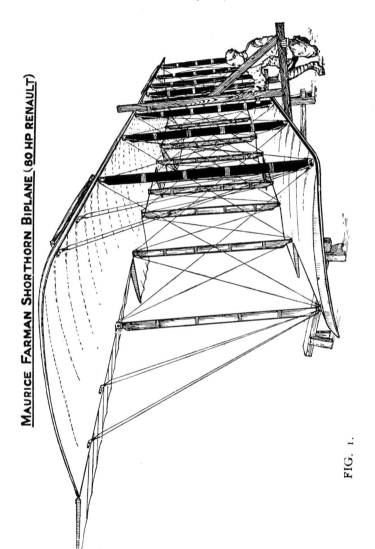

MAURICE FARMAN SHORTHORN BIPLANE (80 HP RENAULT)

FIG. 1.

T.5. 1/18.

MAURICE FARMAN "SHORTHORN" BIPLANE

(Type 1914).01   80 H.P. Renault.

Temporary diagonal wires
front and rear

Strut A front rear	Strut B front rear	Strut C front rear	Strut D front rear	Strut E front rear	Strut F front rear	Strut G front rear	Strut H front rear

5½"   5½"

Fig. 1
FRONT ELEVATION OF MAIN CELL.

Erection and Preliminary Truing Up of Main Cell
Erect Lower Main Plane complete less Ailerons, on low trestles placed under
Struts A.T. C.R. F.M. H.K.
Bolt up the Upper Main Plane complete and fit Struts A.T.C.R.F.M.H.K. into their
Top Sockets By means of these Struts lift the Upper Main Plane bodily
into position fitting the bottom ends of the Struts into the Sockets on
the Lower Main Plane.
Fit all remaining Struts and loosely connect all Flying, Landing and
Incidence Wires.
Connect four temporary diagonal wires, two in front and two in rear
at Struts D. E. N. P.
True up Centre Section completely. (see fig 2)
Check by stretching a line from tops of Bottom Strut Sockets at C and
F This line must be 5 millimetres below the tops of Sockets at
bottom of Struts D and E.
By means of a large square adjust so that Centre Section Struts are
vertical (see fig 2) The lower limb of the square must touch the
under side of Front Spar and Trailing Edge just outside Centre Section
and must be horizontal.
Next by means of the Front Flying and Landing Wires adjust the
Leading Edge of Lower Main Plane so that a line stretched from tops
of Sockets at bottom ends of Struts A and H cuts the tops of
Sockets at bottom ends of Struts B. C. F. G. and is 5 millimetres below
tops of Sockets at bottom ends of Struts D and E.

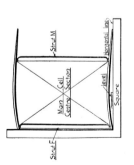

Fig. 2

G Sheets
Sheet 1

T.; 1/18.
MAURICE FARMAN "SHORTHORN" BIPLANE
(TYPE 1914).   80 H.P. RENAULT.

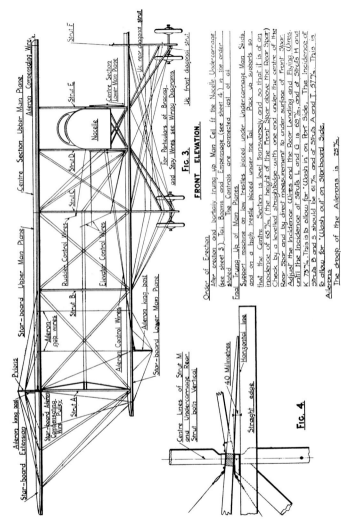

FRONT ELEVATION.

FIG. 3.

FIG. 4.

Order of Erection

After erection and partially truing up Main Cell, fit the Nacelle Undercarriage (see sheet 3). Tail Booms and Empennage (see sheet 4) in the order stated. The Controls are connected last of all.

Final truing up of Main Planes.

Support machine on low trestles placed under Undercarriage Main Skids and on a high trestle placed under the Tail. Pack up supports so that the Centre Section is level transversely and so that it is at an incidence of 65%o (the height of the Front Spar above the centre of the Rear Spar and by direct measurement to under surface of Front Spar. Check by a levelled straightedge with one end under the centre of the Rear Spar. Adjust the Incidence Wires and the Rear Landing and Flying Wires until the Incidence at Struts L and G is 69%o and at Struts H and K 75%o. This is to allow for 'Wash in' on Port Side. The Incidence at Struts B and S should be 61%o and at Struts A and T 57%o. This is to allow for 'Wash out' on Starboard Side.

Ailerons.

The droop of the Ailerons is 25%o.

T.5. 1/18.

MAURICE FARMAN "SHORTHORN" BIPLANE

(Type 1914). 80 H.P. Renault.

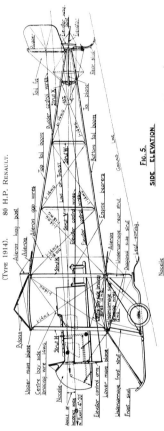

Fig. 5.

SIDE ELEVATION.

Nacelle.
To fit Nacelle tie ropes over Nacelle supports on Struts D.E.N.P. (see fig 1.) to Bottom Ends of Struts C.F.M.R. respectively, thus slightly expanding middle bay of Centre Section to allow Nacelle to be placed in position without damaging the Struts D.E.N.P.
Lift Nacelle in position and bolt up to Struts D.E.N.P. and to Cross Tubes front and rear.

Undercarriage.
Before fitting Undercarriage lift the Main Planes on to both trestles placed under Struts A.T. and H.K. and pack up until Centre Section Struts are vertical (see fig 6.)
Undercarriage Rear Struts must be vertical and in line with Struts M and R.
Undercarriage Front and Rear Struts must be in one and vertical when viewed from front.
Measurement from the Centre at Bottom End of Undercarriage Front Strut to plumb line touching the Leading Edges of both Main Planes must be 10¾ ins as shown in fig 6.
Adjust by Undercarriage Bracing Wires.

Flying Position.
The Machine is in a flying Position when the Engine Bearers and Longerons are horizontal. Check by short level on Engine Bearers.

Fig. 6.

6 Sheets
Sheet No. 1

A

MAURICE FARMAN "SHORTHORN" BIPLANE

(Type 1914).

80 H.P. Renault.

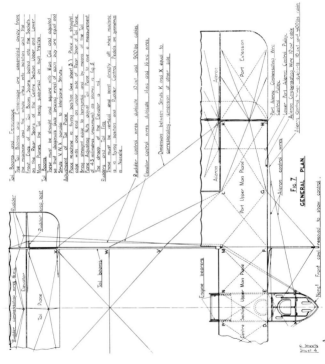

Tail Booms and Empennage.
The Tail Booms and Empennage are assembled away from the machine and the whole idea into position until the Front Ends of the Tail Boom Booms pushed into the Sockets on the Rear Spars of the Centre Section Upper and Lower Main Planes the Tail being supported on high trestles.

Tail Booms.
These must be straight and square with Main Cell and adjusted so that diagonal slack bracing wires of each bay are equal and square V.X.X. parallel to Interplane Struts.

Adjustment of Tail Plane.
Place machine in flying position (see sheet 3). Place a straight edge with one end on under side of Rear Spar of Tail Plane. Bring straight edge to horizontal and by means of the Tail Plane Adjusting Nuts adjust Tail Boom to give a measurement of 43 millimetres (maximum) as shown in fig 5.

The droop of the Elevator is nil.

Rudders and Tail Fins.
These must be vertical and point directly aft when machine is in flying position and Rudder Control Pedals are generally a Nacelle.

Rudder control wires duplicate 10 cwt and 900 lbs cables

Elevator control wires duplicate 7½ws and 16 ws wires

Dimension between Struts K and X equal to corresponding dimension of other side

FIG. 7
GENERAL PLAN

Labels on drawing: Rudder · Elevator · Rudder wing post · Tail Plane · Rudder compensating wires fixing · Tail booms · W · Y · Engine bearers · Centre Section Upper Main Plane · Port Upper Main Plane · Port Extension · Aileron · Aileron · Aileron · Port Aileron Compensating Wire · Gaithor Pulley · Lower Port Aileron Control Pulley · Aileron Compensating Wire 10 cwt cable · Aileron control wires · Aileron Control Wires duplicate 10 cwt and 900 lbs wire · Port Upper Main Plane · Port Aileron Compensating Wire · Nacel Front cowl removed to show control.

T.5. 1/18.

MAURICE FARMAN "SHORTHORN" BIPLANE

(Type 1914).   80 H.P. Renault.

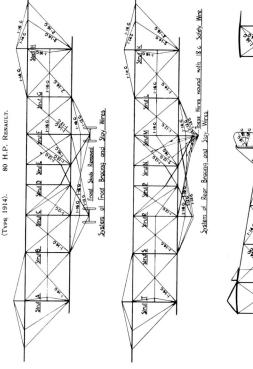

System of Front Bracing and Stay Wires

System of Rear Bracing and Stay Wires

These Wires wound with 18 g. Safety Wire

System of Side Bracing and Stay Wires

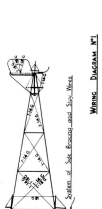

Side Bracing Wires at Struts F, N and D.P.

WIRING DIAGRAM Nº1.

6 Sheets
Sheet 5

A

T.5; 1/18.

MAURICE FARMAN "SHORTHORN" BIPLANE

(Type 1914). 80 H.P. Renault.

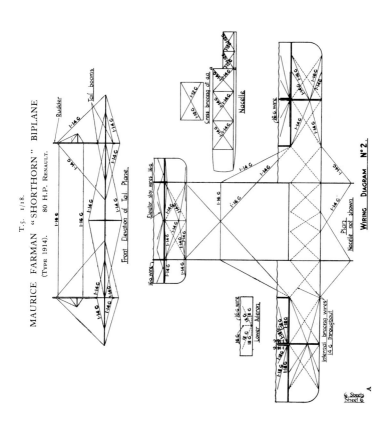

Front Elevation of Tail Plane.

Rudder

Tail booms

Elevator stay wires I6g.

Cross bracing at a a.

Nacelle

Lower Aileron

Plan.
Nacelle not shown.

Internal bracing wires 14 G throughout.

WIRING DIAGRAM Nº 2.

6 Sheets
Sheet 6

A

NOTES.

AIRBOARD TECHNICAL NOTES

SOPWITH BIPLANE F.1.

(130 H.P. CLERGET.)

RIGGING NOTES.

SOPWITH BIPLANE F. 1.

(130 H.P. CLERGET.)

Camel

MANUFACTURERS' ORDER OF ERECTION.

1. Fuselage assembled and trued up.
2. Tail Skid fitted and Cloche fitted.
3. Reinforcements of Top Longerons in Pilot's Cockpit fitted.
4. Fairing, Dashboard and Deck fitted.
5. Controls fitted in front of Fuselage.
6. Undercarriage fitted.
7. Machine erected in Flying Position with Main Planes and Empennage in skeleton and trued up.
8. Controls loosely connected.
9. Main Planes and Empennage dismantled.
10. Petrol Tanks fitted.
11. Fuselage covered and doped.
12. Engine Mounted and Engine Accessories and Controls fitted.
13. Gun Magazine and Shutes fitted.

Sopwith Biplane F. 1 (130 H.P. Clerget).

14. Press Plate at bottom of first bay fitted.

15. Oil Tank fitted.

16. Gun and Actuating Gear fitted.

17. Centre Section fitted (after being covered and doped).

18. Wheels and Shock Absorbers fitted on Under-carriage.

19. Cowling fitted.

20. Main Planes and Empennage fitted (after being covered and doped.

21. Controls connected up and adjusted.

SOPWITH BIPLANE F. 1.
(130 H.P. CLERGET.)

TRUING UP FUSELAGE (see Fig. 1).

The Side Struts are numbered from front to rear of Machine. Support the Fuselage on two trestles, one placed under the first bay and the other under the last bay.

Starting at No. 1 vertical Side Strut mark points on consecutive Vertical Side Struts $15\frac{1}{4}''$, $15\frac{1}{4}''$, $15\frac{1}{4}''$, $15\frac{1}{4}''$, $15\frac{1}{16}''$, $13\frac{17}{32}''$, $11\frac{5}{16}''$, $9\frac{1}{2}''$, $7\frac{25}{32}''$, respectively below the *upper surface* of the Top Longerons. These marked points must be along the Thrust Line in Side elevation when Fuselage is trued up.

Lightly clamp a straightedge tranversely across No. 3 Side Struts, the marked points to be on the *upper edge*.

Lightly clamp small blocks of wood at all other marked points, the marks to be on the *upper edge* of the blocks.

Proceed to true up as follows :

Tension Internal Cross Bracing Wires until diagonals are equal in each section. Check by trammel.

Make Top Cross Bracing Wires equal in each bay and similarly make Bottom Cross Bracing Wires equal in each bay. Check by trammel.

Level the transverse straightedge by suitably packing up the front support.

Adjust the Side Bracing Wires on one side until any two consecutive marked points on Vertical Side Struts are in a horizontal line and all Struts are vertical.

Check the former by placing a straightedge across consecutive blocks and adjust until the straightedge is level and check the latter by plumbing each Strut.

Proceed similarly on the other side.

The Sternpost should be vertical viewed from any direction when the Fuselage is trued up.

TRUING UP UNDERCARRIAGE (see Fig. 2).

The Undercarriage is trued up by making Diagonal Bracing Wires equal in length. Check by trammel.

PLACING MACHINE IN FLYING POSITION.

Before truing up the Centre Section and fitting the Main Planes it is necessary to get the Machine in Flying Position.

NOTE.—The "Figs." refer to the Illustrations and *not* to the Rigging Diagrams.

Sopwith Biplane F. 1 (130 H.P. Clerget).

To do this support the Machine by blocks placed under the Undercarriage Struts and on a trestle placed under the Tail. The Machine is in Flying Position when the Front Spar at the Bottom of the Fuselage to which the Front Spars of the Lower Main Planes are attached, is level transversely and when the Top Longerons in the Pilot's Cockpit are level longitudinally. Level longitudinally by raising or lowering Tail and transversely by packing blocks under the Undercarriage Struts.

TRUING UP THE CENTRE SECTION (See Figs. 2 and 3).

The Centre Section is symmetrical about the vertical centre line of Machine. True up by Centre Section Cross Bracing Wires and ensure that both Upper Wires are equal in length and also that both Lower Wires are equal in length.

The *Stagger* of the Centre Section is 18". This can be adjusted by the Stagger Wires *after* the Lower Main Planes have been fitted.

Check by dropping plumb lines from the Leading Edges of the Centre Section ; the *fore* and *aft* horizontal distance of the Leading Edge of the Lower Main Planes should be 18" from the plumb lines.

ATTACHING THE MAIN PLANES.

Place the Lower Main Planes in position and insert the Securing Rod on each side from the *rear*.

Push the latter Rod well in and secure it to the *Ribs* at the Roots of the Main Planes by nuts and bolts, not forgetting to insert the *Split Pins*.

Loosely connect the Landing Wires.

Lift the Upper Main Planes in position, insert the Securing Rods from the *front* and after pushing the Rods well in, secure the front of the Rods to the Leading Edges of the Main Planes.

Loosely connect Flying and Incidence Wires.

TRUING UP MAIN PLANES.

Drop plumb lines from four points, two each side, on the Leading Edge of the Upper Main Planes. True up until—

(*a*) The plumb lines are in line viewed from the side.

(*b*) The Leading Edge of the Upper Main Planes is straight viewed from the front and in plan view, there being no *Dihedral* on the Upper Main Planes.

(*c*) The Leading Edge of the Lower Main Planes is symmetrical about centre line of Machine.

Check by taking measurements from Bottom Sockets of Front Outer Struts to Rudderpost and Propeller Bos Corresponding measurements should be the same on both sides.

(*a*) The *Dihedral* of the Lower Main Planes is 5°. Check by Abney level and straightedge along the Front and Rear Spars.

5

(e) The *Stagger* is 18″ at *Centre Section* and 18$\frac{5}{16}$″ at *Outer Struts.* Check by measuring the *fore* and *aft* horizontal distance from the Leading Edge of the Lower Main Planes to the plumb lines dropped from the Leading Edge of the Upper Main Planes.

(f) The *Incidence* is 2° throughout. Check by Abney level and straightedge, taking care to place the latter from Leading Edge to Trailing Edge at *Ribs.*

(g) There is no "*Wash In*" or "*Wash Out.*"

FIXING THE EMPENNAGE. (See Figs. 3 and 4).

Bolt the Tail Plane in position and connect the Tail Plane Bracing Wires

True up the Tail Plane to be level transverely.

Check by spirit level along the Front Spar and along the Hinged Tube.

Check for Tail Plane being square with Machine by taking measurements from Bottom Sockets of Rear Outer Struts to lateral extremities of Rear Spar of Tail Plane.

These measurements should be the same on both sides.

The *Incidence* of the Tail Plane is 1$\frac{1}{2}$°. Check by using a straightedge with two small blocks attached. These blocks should be of such dimensions that when one is on the Front Spar and the other on the Hinged Tube of the Tail Plane, the *upper edge* of the straightedge is parallel to the *fore* and *aft* centre line of the Tail Plane.

The *Incidence* can be measured by an Abney level over the straightedge with Machine in flying Position.

Bolt the Fin in Position.

Hinge the Rudder to the Sternpost and Fin, not forgetting to insert the Split Pins, similarly hinge the Elevators to the Rear Spar of Tail Plane.

NOTE.—The two Elevators should be rigidly connected, so that if the Elevator Control is damaged on one side the Elevators can be operated from the other side.

CONTROLS.

Connect up the Controls and adjust them so that

(a) With the Pilot's Control Stick central there is no *Droop* on the Ailerons.

(b) When the upper *front edge* of the tubular reinforcement on the Pilots Control Stick just beneath the handle is 9$\frac{3}{4}$″ *horizontally in rear* of the Dashboard, the Elevators are in continuation of the Tail Plane.

(c) The Rudder and Tail Skid point directly *fore* and *aft* and are square with Machine when the Rudder Bar is square in Fuselage.

SOPWITH   BIPLANE   F.   1.

(130   H.P.   CLERGET.)

LIST   OF   PRINCIPAL   DIMENSIONS.

Span of Upper Main Planes } „ Lower „ „	28′ 0″
Chord of Upper Main Planes } „ Lower „ „	4′ 6″
Incidence of Upper Main Planes } „ Lower „ „	2°
Gap at Fuselage -	5′ 0″
Stagger at Centre Section -	18″
„ Outer Struts -	18 $\frac{5}{16}$″
Dihedral of Upper Main Planes	Nil
„ Lower „ „ -	5°
Overall Length	18′ 9″
Height	8′ 6′
Incidence of Tail Plane (Centre Line)	1$\frac{1}{2}$°
Droop of Ailerons-	Nil

Elevators are in continuation of Tail Plane when Upper Front Edge of Tubular Reinforcement on Pilot's Control Stick, just beneath the handle is 9$\frac{3}{4}$″ horizontally in rear of Dashboard.

Sopwith Biplane F. 1 (130 H.P. Clerget).

POINTS TO OBSERVE WHEN OVERHAULING MACHINE.

See that the Leading Edges of the Main Planes are symmetrical, about centre line of Machine.

Examine the Bracing Wires for length and tautness in the Centre Section, and see that the Split Pins are in position, and that all Lock Nuts are tight.

Check the Dihedral of Lower Main Planes.

Check the Stagger.

Check the Incidence.

See that the Interplane Struts are straight.

Examine all Main Plane Bracing Wires for length and tautness, and see that all Split Pins are in position.

Examine all Controls, Control Pulleys and Cables, see that they work freely and that Turnbuckles on Cables are locked.

Examine Tail Plane and see that it is set correctly and is square with Machine, and that all Tail Plane Bracing Wires are correct both as to tautness and length, and that all Split Pins are in position, and that all Lock Nuts are tight.

Examine Rudder and Fin and see that they are set straight and square with Machine.

Measure the Droop of the Ailerons and Elevators.

Examine Undercarriage and Skid.

Examine Tank Mountings and Connections.

Examine Engine Mounting, Engine Controls, and Engine Accessories.

Examine Cartridge Drums and see that they are secure and do not foul the Carburetter.

A.B.T.D., T5., 1/18. SOPWITH BIPLANE F.1. (130 H.P. Clerget.)

FIG. 1.

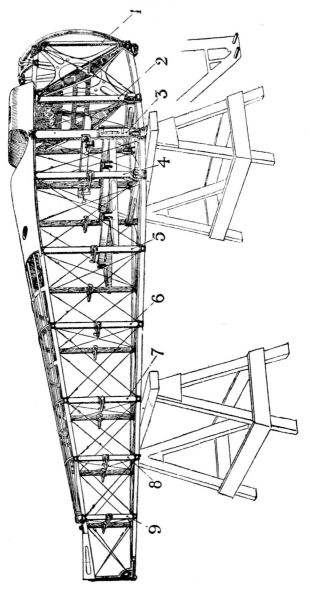

A.B.T.D., T.5., 1/18.   SOPWITH BIPLANE F.1.   (130 H.P. Clerget.)

FIG 2.

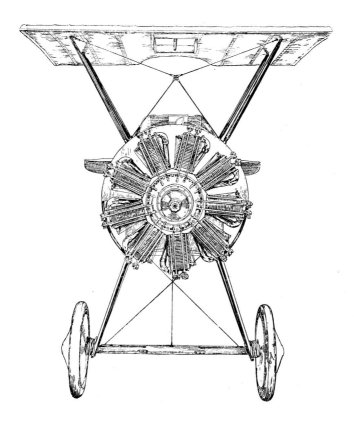

NOTE.—This illustration shews the 150 H.P. A.R.1., which
in some cases is fitted.

A.B.T.D., T.5., 1/18. SOPWITH BIPLANE F.1. (130 H.P. Clerget.)
FIG 3.

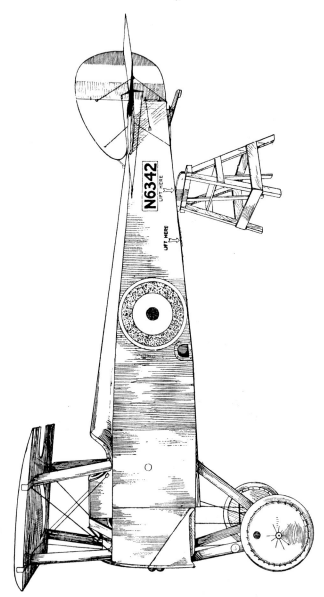

A.B.T.D. T5. 1/18.   SOPWITH BIPLANE F.1.   (130 H.P. Clerget.)
FIG. 4.

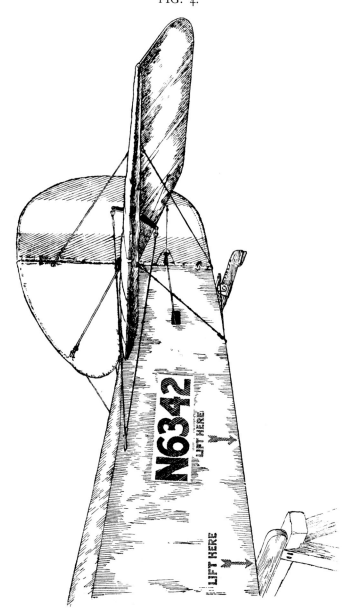

AIRBOARD   TECHNICAL   NOTES

SOPWITH   BIPLANE   F.1.

(130   H.P.   CLERGET.)

RIGGING   DIAGRAMS.

SOPWITH BIPLANE. F.I.

130 H.P. CLERGET.

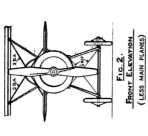

FIG. 1.

SIDE ELEVATION

Elevator Control 1.15 Cut. Cable.
Skid Control N.G. Piano Wire.
Aileron Wire Control 1.15 Cut. Cable.
Tail Skid.
Rudder Control 2.15 Cut Cable.
overall 18'-9" approx.

FIG. 2.

FRONT ELEVATION

(LESS MAIN PLANES)

Flying Position:
To set Machine in Flying Position, level transversely across the lower Front Spar in Fuselage and longitudinally across the upper surface of the Top Longerons in the Pilot's Cockpit.

Trueing up Fuselage:
The Fuselage is symmetrical in plan view and from Strut No.6 to Rudder Post in Side Elevation.
Mark points on Side Elevation so that these points are on the Thrust Line in Side Elevation (for distances below the upper face of the Top Longerons see Table A Sheet 4.)
Make Internal Cross Bracing Wires equal at each section. Check by Trammel.
Make Top Cross Bracing Wires equal in each Bay. Similarly make Bottom Cross Bracing Wires equal in each Bay. Check by Trammel. True up by Side Bracing Wires until all marked points on Side Struts are in line in Side Elevation, levelling from marked point to marked point on adjacent Side Strut and checking each Side Strut for being Vertical.

Trueing up Undercarriage:
True up until Cross Bracing Wires are equal. Check by Trammel.

Trueing up Centre Section:
The Centre Section is symmetrical about the Vertical centre line of Machine. Adjust by Centre Section Cross Bracing Wires until Upper Wires are equal and Lower Wires are equal. Check by Trammel.
The Stagger of the Centre Section is 1.6 Check after fitting Lower Main Planes and adjust by Centre Section Stagger Wires.

SOPWITH BIPLANE. F.I.

130 H.P. CLERGET.

Fig 3
FRONT ELEVATION

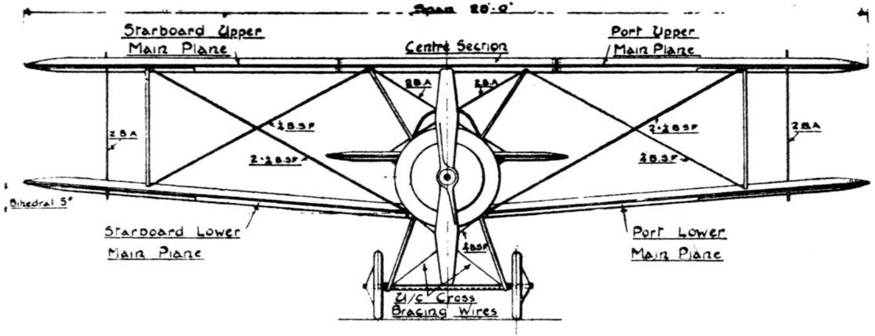

Span 26'9"

Starboard Upper Main Plane
Starboard Lower Main Plane
Port Upper Main Plane
Port Lower Main Plane
Centre Section
B/C Cross Bracing Wires

Dihedral 5°

POSITION OF UPPER MAIN PLANE IN RELATION TO LOWER MAIN PLANE

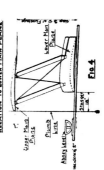

Fig 4

Upper Main Plane
Lower Main Plane
Plumb Line
Abney Level
Incidence 2°
Stagger 1¾"
Gap 5'3¾"

4 Struts
Sheet 8.

MAIN PLANES

The Leading Edge of Upper Main Planes is straight in plan view & viewed from front. The Leading Edge of Lower Main Planes is symmetrical about centre line of Machine. Drop Plumb lines on each side from the Leading Edge of the Upper Main Planes. These lines must be in line viewed from the side.

Dihedral
The Dihedral of the Lower Main Planes - 5° Check by Abney Level and Straightedge along the Spars. There is no Dihedral on the Upper Main Planes.

Stagger The Stagger is 18" at Centre Section and 18¾" at Outer Struts. Check by measuring the horizontal fore and aft distances between the Plumb lines and the Leading Edge of Lower Main Planes.

Incidence
The Incidence is 2° throughout Upper & Lower Main Planes. Check by Straightedge and Abney Level from Leading Edge to Trailing Edge at Ribs.

Ailerons
There is no Droop on the Ailerons with Pilot's Control Stick Central.

Sopwith Biplane. F.I.

130 HP. CLERGET.

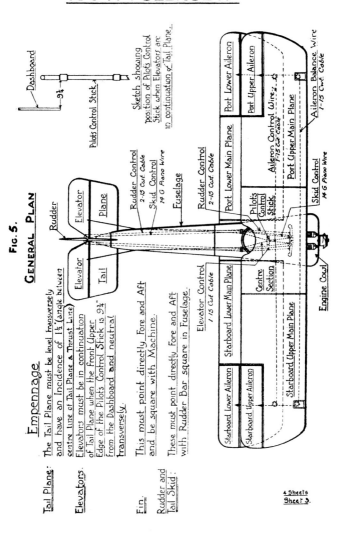

Fig. 5.

GENERAL PLAN

NOTES.

AIRBOARD   TECHNICAL   NOTES

SPAD   BIPLANE,   TYPE   S.   VII.

150   H.P.   HISPANO-SUIZA.

RIGGING   NOTES.

SPAD BIPLANE.

TYPE S VII.

150-H.P. HISPANO-SUIZA.

MANUFACTURERS' ORDER OF ERECTION.

---

1. Fuselage assembled and trued up. Tail **Skid** fitted.

2. Main Petrol Tank fitted with Fuselage in **Inverted** Position.

3. Undercarriage fitted and trued up.

4. Engine mounted with Machine in Flying Position.

5. Cowl fitted.

6. Main Planes attached with Machine in **Flying** Position (Lower Main Planes first).

7. Controls connected up and adjusted.

8. Fuselage covered and doped.

SPAD BIPLANE, TYPE S. VII.

(150 H.P. HISPANO-SUIZA).

TRUING UP FUSELAGE (See Fig. 1).

The Fuselage Struts are numbered from front to rear of Machine.

Place the Fuselage with Top Longerons uppermost on two trestles, one between Struts 1 and 2 and the other under the last bay.

Pack up with blocks so as to level the Top Longerons as far as possible.

Mark points on Struts 2, 4, 6, 8 on both sides of Fuselage 9in. below *top* face of Top Longerons.

Lightly clamp straightedges transversely between corresponding marked Struts, the marked points to be on the upper edges.

Mark middle points of all Cross Struts, and mark another lot of points half an inch to one and the same side of all these points.

By means of jigs fitting over the Rudderpost and Engine Bearers (see sketch on Page 11) the *true centre line* of Machine is found. From this a line is stretched half an inch *out of centre* on the *same* side as the second lot of marked points on the Cross Struts, the object being to prevent the plumb lines fouling the Fairing Stringers.

Then by means of the Fuselage Bracing Wires true up until all straightedges are level transversely.

Adjust for longitudinal level also, using a long straightedge, over the faces of the transverse straightedges for this purpose.

Plumb lines from the second marked point on any Top Cross Strut should cut the corresponding point on the Bottom Cross Strut and the line half an inch out of centre.

TRUING UP UNDERCARRIAGE.

By means of the front and rear Cross Bracing Wires true up the Undercarriage so that corresponding diagonals are equal. Check by trammel. (See Figs. 2 and 5.)

PLACING MACHINE IN FLYING POSITION.

Before truing up Centre Section and attaching the Main Planes it is necessary to get the Machine in Flying Position.

To do this block the Machine up under the Undercarriage Struts and support the Tail by a trestle placed near the Rudderpost.

Mark points on Struts 2 and 3 on both sides of Fuselage at equal distances below the *under* surface of the Top Longerons.

NOTE.—The "Figs." refer to the illustrations and *not* to the Rigging Diagrams.

Spad Biplane, Type S VII. (150 H.P. Hispano-Suiza).

Fix a straightedge longitudinally to the marks on each side and level by raising or lowering Tail.

Level transversely by packing blocks under the Undercarriage Struts.

TRUING UP CENTRE SECTION.

By means of the Adjusting Steel Tubes and side Cross Bracing Wires true up the Centre Section so that the Struts are vertical when viewed from the front, and so that the Sockets on tops of the Struts will correspond to the fittings on the Upper Main Plane; check by trammel. (See Figs. 2 and 3).

A plumb line from the centre of Top Socket on Front Centre Section Strut must pass across the centre of the hole in the Bottom Longeron into which the Front Spar of Lower Main fits; check both sides. (See Sketch, Page 10 and Fig. 5.) Similarly for the Rear Centre Section Struts.

N.B.—The Centre Section should be trued up *before* mounting the Upper Main Plane.

FITTING LOWER MAIN PLANES.

See that the Lower Main Planes have the Bell Crank Aileron Control Levers properly bolted in position.

See that the Rods connecting Pilot's Control Stick and Bell Crank Levers are loosely inserted in the Lower Main Planes, and are not bent or damaged.

Place the Lower main Planes in position, inserting the projecting portions of the Spars at the root into the recesses in the Bottom Longerons, also passing the horizontal Aileron Connecting Rods through the holes in the Bottom Longerons just behind the Rear Spar holes (see Fig. 5), but not securing them either to Pilot's Control Stick or Bell Crank Lever until the Planes have been trued up.

Support the Planes near the Wing Tips on trestles.

Join up the Piano Wires at the *bottom* of the Fuselage which connect the Spars of the Lower Main Planes.

INTERPLANE STRUTS AND FITTINGS.

The two halves of the Intermediate Struts, Front and Rear, the horizontal connecting Strut, and the cables connected to the fittings to which all are attached are connected up away from the Machine.

Fit all the Struts in position on the Lower Main Planes. (See Fig. 6.)

The Struts are secured to the Wiring Plates on the Planes by Pins, which are prevented from coming out by ordinary domestic Safety Pins, and *not* by *Split Pins*.

Spad Biplane, Type S VII. (150 H.P. Hispano-Suiza).

FITTING THE UPPER MAIN PLANE.

The Upper Main Plane is lifted into position and bolted to the Sockets on the tops of the Centre Section Struts.

The *Bolts* pass from the *under surface* to the *upper surface,* the *Nuts* being on *top. Spring Washers* are fitted *under* the *Nuts.*

The Intermediate and Outer Struts are fitted to the Upper Main Plane in the same way as they are to the Lower Main Planes, i.e., by Pins passing through the Fork End Fittings, and secured by *Safety Pins.*

Loosely connect up all Landing, Flying, and Incidence Wires.

After mounting the Upper Main Plane in position connect the Leads to the Petrol and Water Tanks in the Centre Section of the Upper Main Plane.

IMPORTANT.—The Bolts securing the Strut Fittings on the Upper Main Plane pass from *under surface* to *upper surface,* and those securing Strut Fittings to the Lower Main Planes from *top surface* to *under surface.*

TRUING UP MAIN PLANES.

All Interplane Struts must be vertical when the Machine is in Flying Position.

There is no *Dihedral.*

The *Stagger* is 45mm.*

This should be checked by dropping a plumb line from the Leading Edge of the Upper Main Plane; the fore and aft horizontal distance of this line in front of the Leading Edge of the Lower Main Plane should be 45mm.* when the Machine is in Flying Position.

The centre of the fitting to which both halves of the Intermediate Struts are fitted is the mid point of the diagonals along which the Landing and Flying Wires lie. (See Fig. 6.)

Check by trammel from this centre point to centre points on Strut fittings and ensure that these distances are equal.

Leading Edges of Upper and Lower Main Planes should be straight transversely, and at right angles to the longitudinal axis of the Machine.

The *Incidence* of the Lower Main Planes is 1deg. 30min. (1½deg.) throughout.

Check by straightedge and protractor or Abney level, taking care to place the straightedge from Leading Edge to Trailing Edge at *Ribs* of the Lower Main Planes.

The *Incidence* of the Upper Main Plane is 2° throughout. Check as for Lower Main Planes.

There is no " *Wash In* " or " *Wash Out.*"

---

* The Stagger is 45mm. according to design, but in practice it is found that 43mm. is the average value as measured on a large number of Machines.

Spad Biplane, Type S VII. (150 H.P. Hispano-Suiza).

The distance from the centre of *Inner Bolt* at *foot* of **Front Outer Strut** to centre of Propeller Boss should be 10ft. 5⅜in. on both sides.

The distance from the centre of *Inner Bolt* at *foot* of **Front Outer Strut** to centre of *Lower Front Bolt* on side of **Rudder Plate** should be 17ft. 0⅜in. on both sides.

AILERON CONTROLS.

After truing up the Main Planes connect the vertical Aileron Control Rods along the Rear Outer Struts to the Kingpost on the Ailerons and to the Bell Crank Aileron Control Levers, not forgetting the *Split Pins*. Then fit the Fork Ends of the horizontal Aileron Connecting Rods in the Lower Main Planes to the Pilot's Control Stick. The latter rods are then adjusted by turning them until they fit in position in the Bell Crank Levers when the Pilot's Control Stick is central. (See Fig. 7.)

The *Droop* of the Ailerons is a quarter of an inch.

FIXING THE EMPENNAGE.

Pass the Rudder and Elevator Control Wires through the Tail Plane connecting the latter wires to the Kingpost, inside the Tail Plane, which controls the Elevator.

Bolt the Tail Plane to the top of the Fuselage and insert the *Split Pins.* Tail Plane to be packed up 3-16in. in rear (result of tests up to 6/5/17).

Bolt up the two steel Supporting Tubes and adjust them so that the Tail Plane is square with the Machine. (See Fig. 4.)

This can be checked by aligning the Trailing Edge of Upper Main Plane and Rear Spar of Tail Plane.

Bolt the Fin in position.

Hinge the Rudder to the Rudderpost by passing the long Cotter Pin through the *Eye Bolts* on Rudder and Rudderpost, and secure the eye end at the bottom of the Rudder Cotter Pin to the lowest *Eye Bolts* on the Rudder and Rudderpost by wire.

Connect the Rudder Control Wires.

Adjust the Controls so that when the Rudder Control Bar is square with Fuselage the Rudder is pointing directly *fore* and *aft*, and is square with the Machine, and when the Pilot's Control Stick is central, but leaning forward ¾in., there is no *Droop* on the elevators.

SPAD BIPLANE TYPE S. VII.
(150 H.P. HISPANO-SUIZA.)

WATER CIRCULATION.

In order to ensure a *constant pressure of water* over the cylinders, a small Tank is placed in the front part of the Centre Section of the Upper Main Plane.

To obtain the best results, it is necessary to prevent all Air Pockets, and to effect this the following procedure is strongly recommended by the Manufacturers:—

1. Place the Machine in Flying Position.
2. Remove Stopper from top of Radiator.
3. Open Tap or Plug placed in top of Carburetter.
4. Fill up with *distilled* or *rain water* (adding about 1½ gallons of glycerine in Winter through Radiator Filler).
5. Replace Radiator Stopper.
6. Remove small Tank Stopper and open small Tap placed under Tank.
7. Pour water into Tank until it runs out at top of Carburetter, then close Carburetter Tap, and continue to pour water until it begins to run out at Tank Overflow Tap. At this point stop filling (*the Tank should then be about half full*) and close the Tap.
8. Start Engine and let it run for 3 or 4 minutes until temperature rises to about 48° in Summer and 65° in Winter.
9. *Stop Engine.* Open Tap or Plug over Carburetter until water begins to run out, add to Top Tank sufficient water to bring water back to previous level (par. 5).
10. Run Engine, one, two, three or four times, letting air out each time at top of Carburetter and examine level of water in Top Tank each time, until no more water is required.

NOTE:—The Stopper of the Radiator is placed *behind* that of the Oil Filler.

If the foregoing instructions are not properly carried out, the result will be a stoppage of Water Circulation to the Radiator, and will probably be followed by the *bursting* of the Top Tank or some other part of the Water Circulation. The Thermometer might not indicate the rise in temperature if the water had completely run out. If there is not sufficient water, owing to either bad filling up or from leakage, *Steam Pockets* will form, and the temperature will rise very rapidly.

SPAD BIPLANE. TYPE S. VII.
150 H.P. HISPANO-SUIZA.

LIST OF PRINCIPAL DIMENSIONS.

SPAN OF UPPER MAIN PLANE - - -	25' 8"
CHORD OF „ „ „ - - -	4' 7"
INCIDENCE OF „ „ - - -	2°
DIHEDRAL - - - - - - - -	Nil
SPAN OF LOWER MAIN PLANES - - - (approx.)	25'
CHORD „ „ „ - - -	4' 2"
INCIDENCE OF „ „ - - -	$1\frac{1}{2}$°
GAP - - - - - - - - -	3' $8\frac{1}{2}$"
STAGGER - - - - - - - -	45mm.
DROOP OF AILERONS - - - - -	$\frac{1}{4}$"

DROOP OF ELEVATORS (with Pilot's Control
Stick ¾in. forward) - - - - - - NIL

OVERALL LENGTH - - - - - (approx.) 20' $3\frac{1}{2}$"

PB - - - - - - 10' $5\frac{5}{8}$" }	Measurements
BR - - - - - - 17' $0\frac{1}{8}$" }	(approx.)

where:—

    P is CENTRE of PROPELLER BOSS.

    B is CENTRE of BOLT at foot of FRONT OUTER
        INTERPLANE STRUT.

    R is CENTRE of LOWER FRONT BOLT on side of
        RUDDER PLATE.

SPAD BIPLANE.

Type S VII.

150-H.P. HISPANO SUIZA.

POINTS TO OBSERVE WHEN OVERHAULING MACHINE.

See that Leading Edges of Main Planes are straight in plan view.

Examine all Bracing Wires for length and tautness in the Centre Section, and see that Split Pins are in position.

Check Planes for level (no Dihedral).

Check Planes for Overhang (the term "Overhang" is used in preference to Stagger, as the Interplane Struts are vertical and Overhang is caused by Lower Main Planes having a smaller Chord than Upper Main Plane).

Check Planes for Incidence.

See that Interplane Struts are straight, and that Safety Pins are in position.

Examine all Main Plane Bracing Wires for length and tautness and see that all Split Pins are in position.

Examine all Cables and Controls and see that same work freely and that all Turnbuckles on Cables are locked.

Examine Tail Plane and see that it is correctly set and is square with Machine and that all Tail Plane Bracing Stays are correct both as to length and tautness, and that all Split Pins are in position.

Spad Biplane, Type S VII. (150-H.P. Hispano Suiza)

Points to Observe when Overhauling Machine.

(Continued.)

Check Main Planes for square with Fuselage and
    Propeller Boss.

Examine Elevator Control Mechanism and see that it
    works freely.

Examine Rudder and Fin and see that they are set
    straight and square with Machine.

Measure the Droop of the Ailerons.

Examine Undercarriage.

Examine Tank Mounting and Connections.

Examine Engine Mounting, Engine Controls and
    Engine Accessories.

Examine Gun Mounting and Gear and check for
    alignment with Engine, also check Sights for
    alignment with Gun.

A.B.T.D. T5. 1/18. SPAD BIPLANE, TYPE S.VII.
(150 H.P. Hispano-Suiza).

TRUING UP CENTRE SECTION.

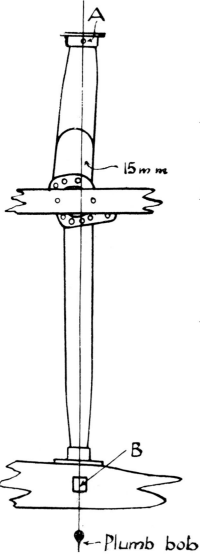

A.—Centre of Bolt on Front Strut Upper Socket.

B.—Centre of recess in Bottom Longeron for Lower Plane Front Spar.

A B must be vertical when Machine is in Flying Position.

OBTAINING TRUE CENTRE LINE OF FUSELAGE.

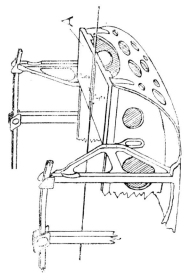

A.—Board fitting accurately between Engine Bearers and marked with true centre line and one parallel line on each side ¾ in. from centre line.

B.—Board with hole accurately fitting Rudderpost and Resting on Top Longerons, marked with true centre line and one parallel line on each side ½ in. from centre line.

A.B.T.D.T5. 1/18. SPAD BIPLANE, TYPE S. VII. (150 H.P. Hispano-Suiza)
FIG. 1.

A.B.T.D.T5. 1/18.   SPAD BIPLANE. TYPE S. VII. (150 H.P. Hispano-Suiza).
FIG. 2.

A.B.T.D.T5. 1/18. SPAD BIPLANE, TYPE S. VII. (150 H.P. Hispano Suiza).
FIG. 3.

A.B.T.D.T5.   1/18.   SPAD BIPLANE TYPE S. VII. (150 H.P. Hispano-Suiza).
FIG. 4.

A.B.T.D.T5. 1/18. SPAD BIPLANE, TYPE S. VII. (150 H.P. Hispano-Suiza)
FIG. 5.

A.B.T.D.T5. 1/18.   SPAD BIPLANE. TYPE S. VII. (150 H.P. Hispano-Suiza).
FIG. 6.

FIG. 6.

A.B.T.D.T5. 1/18. SPAD BIPLANE, TYPE S. VII. (150 H.P. Hispano-Suiza)
FIG. 7.

AIRBOARD TECHNICAL NOTES

SPAD BIPLANE, TYPE S. VII.

150 H.P. HISPANO-SUIZA.

RIGGING DIAGRAMS.

SPAD BIPLANE (TYPE S.VII)
150. H.P. HISPANO-SUIZA.

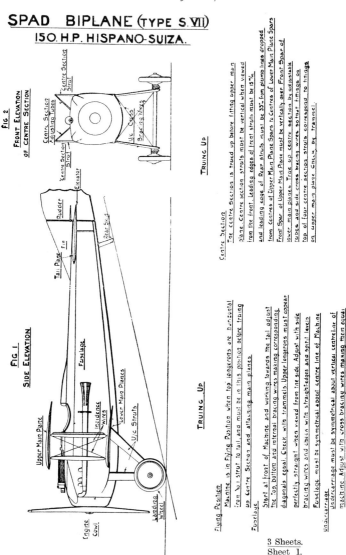

FIG 2
FRONT ELEVATION of CENTRE SECTION

Centre Section Strut

Centre Section Adjusting Tubes

Centre Section Strut

U/c Cross Bracing Wires

TRUING UP

Centre Section

The centre section is trued up before fitting upper main plane. Centre section struts must be vertical when viewed from the front. Leading edges of front struts must be 15″, and leading edge of Rear struts must be 35″ from plumb lines dropped from centres of Upper Main Plane Spars to Centre of Lower Main Plane Spars. Front Spar of Upper Main Plane must be vertically over Front Spar of lower main planes. True up centre section by adjustable tubes and side cross bracing wires so that fittings on top of four centre section struts correspond to hinges on upper main plane. Check by trammel.

FIG 1
SIDE ELEVATION

Elevator

Rudder

Rear Strut

Tail Plane   Fin

Fuselage

Upper Main Plane

Incidence Wires

Lower Main Planes

U/c Struts

Engine Cowl

Landing Wheel

TRUING UP

Flying Position

Machine is in Flying Position when top longerons are horizontal from tail strut to tail and must be in this position before truing up Centre Section and attaching main planes.

Fuselage

Start at front of Machine and working towards the tail adjust the top, bottom and internal bracing wires making corresponding diagonals equal. Check with trammels. Upper longerons must appear perfectly straight when viewed from one side. Adjust with side bracing wires and check with straightedges and spirit levels. Fuselage must be symmetrical about centre line of Machine.

Undercarriage

Undercarriage must be symmetrical about vertical centreline of machine. Adjust with cross bracing wires making them equal.

SPAD BIPLANE (TYPE S.VII)
150. H.P. HISPANO-SUIZA.

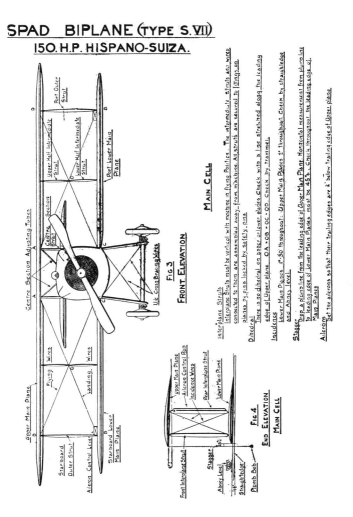

FIG 3
FRONT ELEVATION

FIG 4
MAIN CELL
END ELEVATION

MAIN CELL

Interplane Struts
Interplane Struts must be vertical with machine in flying Position. The intermediate struts and wires connected to them are assembled away from machine. All struts are secured to fittings on plates by pins locked by safety pins.

Dihedral
There is no dihedral on upper or lower plates. Check with a line stretched along the leading edge of Upper plane. OA = OB = OC = OD. Check by trammel.

Incidence
Lower Main Planes 1°30' throughout. Upper Main Planes 2° throughout. Check by straightedge and Abney level.

Stagger
Drop a plumb line from the leading edge of Upper Main Plane. Horizontal measurement from plumb line to leading edge of Lower Main Planes must be 45". Check throughout the leading edge of Main Planes

Ailerons
Set the ailerons so that their trailing edges are ⅛" below trailing edge of Upper plane

A.B.T.D. T5. 1/18.

SPAD BIPLANE (TYPE S. VII)

150 H.P. HISPANO-SUIZA.

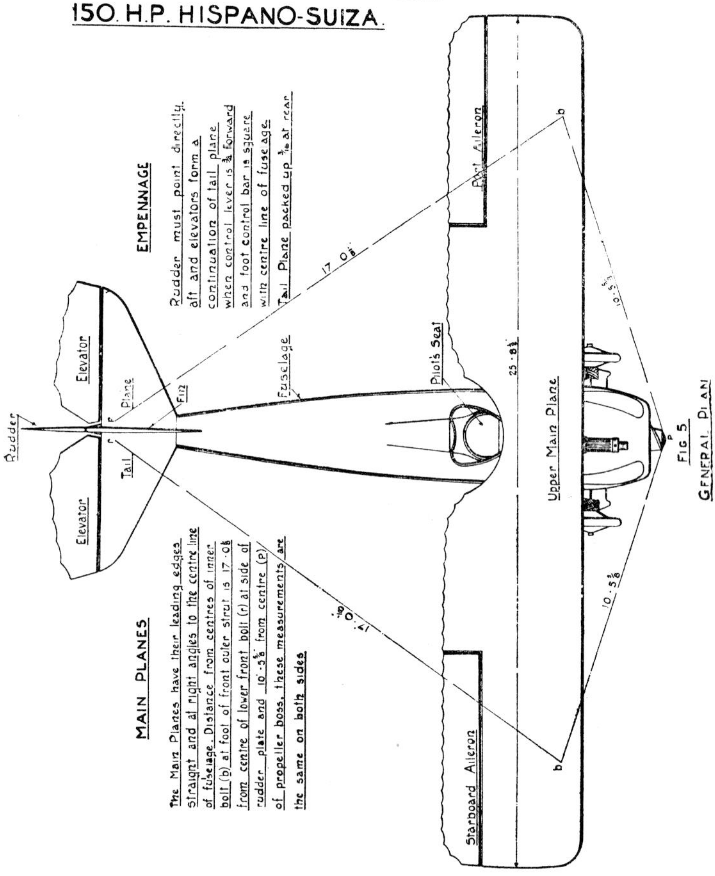

EMPENNAGE
Rudder must point directly aft and elevators form a continuation of tail plane when control lever is ¾ forward and foot control bar is square with centre line of fuselage. Tail Plane packed up 3/16 at rear
MAIN PLANES
The Main Planes have their leading edges straight and at right angles to the centre line of fuselage. Distance from centres of inner bolt (b) at foot of front outer strut is 17·0⅛ from centre of lower front bolt (r) at side of rudder plate and 10·5⅝ from centre (p) of propeller boss, these measurements are the same on both sides
Rudder
Elevator
Elevator
Tail Plane
Fin
Fuselage
Pilot's Seat
Upper Main Plane
Port Aileron
Starboard Aileron
25·8⅛
17·0⅛
17·0⅛
10·5⅝
10·5⅝
b
b
r
p

FIG 5

GENERAL PLAN

3 Sheets

Sheet 3